Learning Diversity

Hans Karl Peterlini

Learning Diversity

University of Klagenfurt (UNESCO Chair on Global Citizenship Education – Cultures of Diversity and Peace – The publication was subsidized by the Klagenfurt University Research Council, University Research Council and University Research Council.)

ISBN 978-3-658-40546-5   ISBN 978-3-658-40547-2 (eBook)
https://doi.org/10.1007/978-3-658-40547-2

Springer VS

Hans Karl Peterlini
University of Klagenfurt
Klagenfurt, Austria

University of Klagenfurt – UNESCO Chair on Global Citizenship Education – Culture of Diversity and Peace – The publication was subjected to a double-blind peer review process. Funded by the Klagenfurt University Research Council and the Faculty of Humanities.

ISBN 978-3-658-40547-2   ISBN 978-3-658-40548-9 (eBook)
https://doi.org/10.1007/978-3-658-40548-9

This Springer VS imprint is published by the registered company Springer Fachmedien Wiesbaden GmbH, part of Springer Nature.
The registered company address is: Abraham-Lincoln-Str. 46, 65189 Wiesbaden, Germany

Contents

Preface: A Weak Pedagogy in a Rough World?

The current living and evolving generations in the so-called Western world have decades of encouraging development behind them: Democracy seemed to be constantly improving in many states. Social balance relativized at least seemingly economic inequality and ethnic-cultural hostility, and global openings also appeared to make the world more convivial and fairer. Within a few years, the world seems to have become gloomier again: The migration and refugee movements from 2014 onward have very quickly weakened international attitudes of solidarity in favor of renationalization of borders. In the measures against the Covid-19 pandemic in 2020, the short-sightedness of primarily national strategies to face global phenomena—such as migration and climate change—became even more manifest. Instead of transnational connectedness and cooperation, which could never be purely economic, massive counter-movements of new nationalist sovereignism are receiving a boost. Along the new divisions dictated by political and economic tactics, new hostilities and threats of war are emerging. Racism is no longer the covert strategy of the parties of yesteryear but a shamelessly used vehicle for inflaming strong populist movements. Fuelled by the election of Donald Trump as US President in 2016, right-wing populist parties are putting even supposedly secure democracies to a risky test, even independently of developments in the US. From West to the East, including Le Pen in France, Salvini, and Meloni in Italy, Orban in Hungary, populist leaders were able to reach a crumbling political center and thereby a majority with ideas once believed to be gone for good:

- National prioritization at the cost of degradation of the Other
- Discrediting of European integration to benefit renationalizing public policies
- Butchering of political language at the expense of careful democratic negotiation processes

H. K. Peterlini, *Learning Diversity*,
https://doi.org/10.1007/978-3-658-40548-9_1

- Retraction of social liberties to strengthen police and military systems
- Private weaponization and legitimacy for violence
- Amplification of national superiority (*"prima gli Italiani,"* "America first")

Simultaneously, there is diminished solidarity for ethnic, cultural, religious, sexual minorities, migrants, and other socially vulnerable people.

This backward movement poses questions we cannot answer solely with contemporary analyses. The fascination with right-wing populist agendas is partly due to a loss of familiarity and security in a globalized world, in which political, economic, social, and climatic developments seem uncontrollable, and seemingly unassailable securities have become unstable. National social fractions, shifts in the global economic balance, threats from climate change, and large migratory movements (due to wars, financial crises, and severe global disarray) lent new strength and credibility to ethnocentric offers of identity.

This dynamic is undoubtedly paradox since the developments described require, on the contrary, transnational strategies, international solidarity, and cooperation in the sense of a global community and planetary politics (cf. Morin and Kern 1999). Climate change and viruses do not stop at national borders, nor can actual social problems be solved by warding off migration and refugee movements. The ethnically, linguistically, socially, ideologically, and religiously plural societies cannot be unified by new homogenization, nor would they gain from it. However, political behaviors and individual strategies for coping with uncertainty and pressure from seemingly unstoppable changes are not necessarily rational. They fall back on patterns and myths that supposedly created unity and averted danger in the past.

The following essays exemplify such attitudes and tendencies on the concept of *Heimat* (*home/homeland*) as needs for shelter and consolidation of both personal and national identity. The articles try to explain, reflect on, and better understand the deeper motives behind the social and psychosocial dynamics, based on the example of young people between ethnocentric belonging and emancipatory development.

In comparison with approaches that seek to overcome cultural identity constructs in favor of intercultural approaches, transcultural changes of perspective (cf. Welsch 1999) or trans-differential refraction (Allolio-Näcke et al. 2005; Breinig and Lösch 2006), the question arises how pedagogy can react to cultural allocations and ethnic hardening. Science and its terminologies can try to liquefy or deconstruct hardened concepts from a critical perspective, such as culture or even *Heimat*. Nevertheless, when dealing specifically with ethnicized people and groups, science will not easily undo the dynamics that guide and promote such allocations and hardenings. In other words, we can view culture as an outdated concept and, as Adorno (1955/2010) astutely recognized, as a substitute for the concept of race that in the German language has become intolerable after the Holocaust. However, this does not mean that people's adherence to a monolithic and narrow understanding of culture will disappear. Cultural open-mindedness cannot be dictated normatively, which would be a contradiction in itself. The following series of articles leads to considerations of how pedagogy can deal with the challenge of ethnic, cultural, and even nationalistic orientations without falling into the trap of a normative and, thus, directive attitude. It sounds like a paradox, but the hope is that answers to the current challenges do not lie in a robust pedagogical approach but in pedagogy that exposes itself to unsecured exchange and for which weakness is its actual strength.

The approach of this book explores pedagogical answers working with the dichotomies at the root of hardening and occlusion. In the context of culture and identity, the focus is on the dichotomy between the I and the Other, amplified in its complexity by a nationally and ethnically defined We against the Others (in plural). Even identity concepts striving for plurality and openness are based on the demarcation of the self from the foreign as a constitutive element. This existential necessity of others to perceive ourselves can hardly be overridden. It is powerful and ultimately unavoidable. But at the same time, it implies that the Other from which I differentiate is existentially necessary for me to consti-tute myself. In "Phenomenology of Spirit" (1807/1976), Hegel conceptualized the parable of bondsman and lord (ibid, pp. 111 ff.), who, despite the submission of the bondsman by the lord, remain trapped in an indissoluble codependence—let-ting unreflected this entanglement bears the danger that the division between the I and the Other is considered absolute cutting the connection. Indeed, if the Other is perceived as so strange and different, not only it *can* be destroyed because it has nothing to do with us, but—depending on the social and political context—it *must* even be eliminated because otherwise it would harass, threaten, infect, or spoil the self or the We. The phantasm of absolute unity and purity of one's own, which must shelter itself from mixing and contamination, is a deadly construct of thought. Unfortunately, its examples are historical and contemporary. They range from the annihilation of Jewish people in the Third Reich to the brutalization in dealing with the weaker and poorer and the cruel, ultimately self-destructive treatment of nature and animals. The drama of the dichotomous division of reality into two pairs of opposites, we consider one half as the norm, while we devalue the other half as a deviation—a seesaw of valency, which is sadly reflected in the dichotomies man-woman, hetero-homo, white-black, mind-body, human-nature, human-animal. These thought patterns have brought us patriarchal

disenfranchisement of women for thousands of years, the persecution of ethnic, cultural, religious, sexual minorities, the blind exploitation of creation, cruel
factory farming, and endangering our biodiversity and health. And they are at
the root of inhumane racism, which in concepts such as culture or ethnicity, if
they are considered biological, only appears a little friendlier, but still prioritizes
division over connectivity. The link between hatred of others and treatment of
nature can be observed in mostly repressed but prominent current developments.
The blind destruction of our livelihoods, the thoughtless, brutal treatment of animals in factory farming, and the extermination of wild animals are evidence that
we still believe we must continue a struggle for survival where reconciliation is
more likely to be required. Even if nature may remain existentially threatening
to humans, we make it frankly more threatening by the delusion of wanting to
control it. The evidence lies in the diseases caused by animal husbandry that is
not appropriate to the species, in the threat to our future food supply through the
impoverishment of biodiversity and, of course, in the simple destruction of the
basis of life through the unrestrained exploitation of our environmental resources.
It is also a matter of finding a balance, a livelihood with the Other because
otherwise, we will destroy ourselves.

What we lose or fight by dichotomous constructions are the nuances, blurs,
and connections between opposing pairs. The dichotomy does not tolerate 'trans'
in the sense of transnational, transcultural, transgender designs and orientations
of people. It only knows the either-or. As a science of the relationship between
people and between humans and the world, Pedagogy has no firm answers to
such challenges. It is not a science of reliable causalities and reliable tips and
tricks. Its answers lie in critical questioning of dichotomies and their formation;
its effort should aim to weaken and unsharpen them in favor of a more sensitive perception of oneself and others. Watching, listening, and feeling, how the
supposedly completely different person appears, *how* they can be met, beyond
stereotyping and abstraction, can reveal the concrete person, the concrete situation, or at least allow perception in a more differentiated way. Such a change of
view would not solve all problems. Still, it does open a way to engage with the
specific person or situation behind the enemy's image, negotiate interests, clarify
positions, and seek solutions.

The awareness for transitions and connections between I and you, We and You,
the Own and the Foreign, between genders and cultures, is not easy and cannot
be prescribed by didactic recipe. It is a matter of delegitimizing a deep-seated
sharpness of division that human orders have drawn over centuries. It requires
the self-reflexive and uncomfortable insight that one's own is always also foreign
and that the foreign always has something to do with oneself. The demand is

for irritating simple belonging and at the same time exposing oneself and other people to more complex contexts—neither a simple nor a reliable path, but it is a walkable one we may dare.

Ultimately, it is about learning for yourself and accompanying others to learn that nothing can exist on their own but that everything and everyone is connected. Empathy is a possible term for retracing this connection. There is no guarantee for saving the world, but it would free those who dare from hardening and occlusion that can neither make them happy nor solve the problems that triggered them.

In this sense, this volume brings together works that have arisen in different contexts (especially Peterlini H.K. 2010a, b, 2011, 2012, 2016a, 2017a, 2019a, b, 2020a),[1]

The book is structured in two major sections, which are themselves subdivided into:

1. The examination of questions of cultural orders that arise as current educational challenges:
   - This chapter deals with *Heimat* as a particularly suggestive projection surface for imaginaries of belonging and, thus, inevitably, of exclusion;
   - For ethnocentric identity concepts, *Heimat* offers itself as a kind of fireplace in which, on the one hand, security is provided and, on the other hand, hardening can take place;
   - By case studies of young people engaged in ethnocentric associations, those understandings of *Heimat* are deconstructed and, by narrative approach, also withdrawn again from a fixative consolidation;
   - For the further development of educational theory and corresponding perspectives for action, the chapter outlines a "pedagogy of relationship" for dealing adequately with dichotomies and stereotypes. The proposed phenomenological approach tries to bracket fixed worldviews, flatter pedagogical hierarchies, and embrace ambivalences. The attitude of looking, listening, and feeling (instead of safe diagnoses and categorizations) goes along with the idea of learning as experience and a narrative approach.
2. Addressing experience as subject and method of a pedagogical response.
   - In the second section, the phenomenological approach is deepened and exemplified in three chapters.

---

[1] These are substantially revised, updated, and written-up contributions, which have been placed in a new context for this volume. The German texts were translated by Dr. Isabel Latz (New Mexico State University) and revised by the author; thanks are also due to Isabella Sandner B.A. for assistance in editing.

- The chapter "The split school" proposes a change of perspective in under-standing normality and claims for sensitization of perception as a necessary pedagogical exercise.
- The chapter "Dialogue with Adorno" reflects the pedagogical approaches discussed on their potential for dealing with xenophobia, racism, and institutional violence.
- The chapter "Searching for the lost Paradis" addresses the dilemma of dichotomous constructions in the perception of the world and the others and reflects the Anthropocene and human interaction with nature and the earth.

As different as they are, the chapters have a common denominator: to understand better how people structure their world, as a basis for exploring possibilities of pedagogical and educational answers. The essays explore whether and, if so, how phenomenological approaches can make dichotomies lighter and more con-scious, even if they cannot surmount their divisive separation. We can hardly abolish dichotomies because we rely on them for our language and our (suc-cessful) thinking habits. Still, we can address the divisive lines of difference and use them for reflections that enable better handling and fraying where security restricts, suffocates, and discriminates. Through the cracks that arise by question-ing the either-or blocks, a glimmer of hope may appear that the "Bildsamkeit," the perfectibility of human beings (as all grand educators have thought), has not come to an end. It's the hope that future generations carry on "what was good," (Peterlini H.K. 2015, p. 37) as the South Tyrolean visionary Alexander Langer wrote in his farewell letter, despite his desperation for the suffering of this world.

The (dis-)order of Culture—Facing Ethnicity and Identity Building

Heimat—Shelter or Cuckoo Nest? Exploration of a Concept Between Belonging and Exclusion

Myth, Magic Formula, Plastic Word—A Conceptual Approach

Heimat is a term that is not only difficult to translate literally but also oscillates between wide-ranging interpretations in German. It can open a wide field or sink into a narrow hole. *Heimat* can be the whole world or your room, an open conglomerate of ideas or a membership card for your party, your football club, or senior citizens' club, a bloated nation or an ethnic minority marginalized by it, such as the great Italy, which bathes in the splendor of antiquity of Rome, or a caravan settlement of Roma or Sinti that is burned down by nationalist fanatics. *Heimat* can be Austria, which shrank after the First World War and still basks in the memory of the Habsburg monarchy in which the sun never set, or a Slovenian village in bilingual Carinthia, which had and has to defend its Slovenian lessons and local plaques against the assimilation pressure of the nation-state. In the election campaign for the office of Austrian President, both the green (and later elected) candidate Alexander van der Bellen (*Standard* 2016) and the right-wing populist candidate Norbert Hofer claimed the proper understanding of *Heimat* for themselves—Van der Bellen with an open perspective, a *Heimat* that can accommodate refugees, Hofer with the closing kind of a *Heimat* that needs to be protected from intruders (cf. *Freiheitsliebe* 2016; *Kurier* 2016).

As such, there is no *Heimat*. Structuralist approaches offer meanings of *Heimat* (cf. Barthes 1972; Deleuze 2003), which can consequently also be interpreted: Substance or truth for the essence of what *Heimat* is, will not be found in these, even with the most thorough psychoanalytic or scientific inspections. However, this does not mean that the different meanings of *Heimat* may not base und on authentic-life situations, facts, circumstances, just as much as on a psychological level the ideas, wishes, fears, constraints, attempts at healing, and hopes connected with *Heimat*.

There is an interplay between myth and truth, which is itself a myth. One creates the other: The myth creates realities—subjectively and inter-subjectively—just as facts, living conditions, living environments, historical and biographical events create myths. According to Jacques Derrida (1973, p. 17), every meaning also plays the role of the significant, the meaning-giving factor.[1] To extrapolate, *Heimat* is not only what we see in it, but it also creates these meanings. The attempt to deconstruct *Heimat* can base on the reciprocity between meaning and the creation of meaning, between meaning and the given circumstances, myth and reality: It is not a search for the real *Heimat*, but an attempt to understand how and why *Heimat* is understood in a certain way.

Stretching a term by Uwe Pörksen (1995), we could see the term *Heimat* as a *plastic word*. It is omnipresent, used indiscriminately and thoughtlessly, and has become a hermetically encapsulated shell, comparable to the drug capsules that people swallow with the confidence that they will help, but whose content remains entirely unexplored for the user. I got this impression when I conducted a series of interviews with young South Tyrolean riflemen (*Schützen*) in 1997/1998 to understand what fascinated them about this traditional association, the sole aim of which was to protect the *Heimat*.

What is *Heimat*? In response to this question, cheeks flushed slightly, mostly there came a "Uhm" and "Whoa," an embarrassed smile, or answers such as "*Heimat*, that's hard to say, that's a lot..." Then attempts to locate it followed: "The place where I live," "The family," "the mountains," "South Tyrol," "Tyrol" (Peterlini H.K. 1998, pp. 76 f.). The deconstruction of *Heimat* is ultimately an attempt to explore the content of this plastic word, in this healing and poisonous pill capsule, not in an essentialist sense, but with the focus on possible meanings.

"*Heimat* – an artificial product for which there was no reality," Walter Jens (1985, p. 15) describes the transformation of a sober word into a romantic attempt to compensate: "*Heimat* as a transfigured yesterday, intact world and relic of the order of the state in the age of urbanization, industrialization, mass collection." (Ibid) In its exaggeration, *Heimat* becomes a counter-world to reality to transfigure it and make it more bearable, according to Freud's understanding of myths as "poetic fancies" (Freud 1922/2011, p. 113) and "the manifestation of distorted cures" (ibid, p. 123), which comfort, but also result in the consolidation of unhealthy and unhappy attitudes. As substitute worlds, myths can prevent

---

[1] Derrida speaks of Sign (understood as meaning in this context) and Signifier (as meaning-giving, meaningful).

problem-solving, as would be possible through communication and negotiation in the real world (cf. Habermas 1984, 1987). According to Mario Erdheim (1984), myths strengthen the rule systems in reality because everything that is conflict-ridden does not give rise to changes in rule relationships but is projected onto external enemies (cf. ibid, p. 38).

The counter-world, the substitute world, or the world of memories as a consolation for the lost: "One must have a home in order not to need it" is the much-quoted formula by Jean Améry (1980, p. 46). Anyone who has lost it, is missing it and looking for it, is exposed to insecurities and disturbances (ibid, p. 47). As the most influential literary *Heimat* factory, German romanticism drew its inspiration from the idealized memory of the time before man's alienation through industrialization and technology.

Similarly, current feelings of disengagement from globalization are trailing the desire for smaller, manageable, protected, and defensive units—that is, *Heimaten*. The plural, which is not yet very common, shows a potential usually reduced to *Heimat* in the singular and thus narrowed to a *Heimat* hermetically delimited by others. The dissolution of borders fosters fear of exposure and loss and creates a longing for safe demarcations and belongings. Thus, *Heimat* becomes a magic that can be invoked against primal fears as well as against current insecurities. These can be worries about the elements, about the cruelty of the Other, about competition and overreaching in an elbow society, about shocks to one's own existence through life events, about the disintegration of one's own ego through psychological stress and threat, about one's own transience and mortality, about political and social processes of change, the effects of which are not yet foreseeable. According to Horst Bienek, such holy words are "used and misused, used and distorted, displaced and exalted, even denounced" (Bienek 1985, p. 7).

The definitions of *Heimat* across the social sciences produce a "confusing series of comparisons and exceptions" (Baur et al. 1998, p. 37). The term receives continued coverage both in the scientific literature and as a cover story in German-language magazines, such as *Der Spiegel* (1984, 2012[2]), *Geo* (2005), and *Du* (2009), in special supplements from renowned daily newspapers such as *Die Presse* (2005), or in high-quality films, for example, the Hunsrück trilogy "*Heimat*" by Edgar Reitz (2010).

---

[2] The cover story of *Der Spiegel* (2012, p. 5) took into account the regional diversification of homeland-feelings in a particular way: The magazine appeared, a unique feature in its history, with 13 different covers, eleven of them for the federal states of Germany, plus separate covers for Austria and Switzerland.

Only a German Issue—or a Universal Need?

Etymologically, *Heimat* is an abstraction from *Heim* (cf. Duden 2007, p. 330). The suffix *-uoti*, which converted *Heim* into *Heimuoti* (Old High German) and *Heimuot* (Middle High German), is the same one that bent the adjective *arm* (poor) into the noun *Armut* (poverty) (ibid). The very concrete word *Heim* for house and place of residence through this suffix became a higher-level, generalizable term. At the linguistic level, just this charging of meaning created the split between everyday practice and overarching meaning, which is also evident in the overlap of *Heimat* in everyday life with political symbols. The word is related to the English *home*, the Swedish *hem*, the Greek *kō̄mē* for village, the Slavic *sem'ja* for family. The noun likely derives from the Indo-European root *kei-* for lying. Possible meanings are the "place where you settle, camp" (ibid), which also includes phrases of *Heirat* (marriage) and *geheuer* (originally part of the household, familiar). A negative dimension of *heim* and *geheuer* reveals a more profound extent that reveals overlaid meanings of security and demarcation: uncanny and *ungeheuer* for what is not your own home, your place, your camp, your own family.

One of the myths surrounding *Heimat* is that *Heimat* is "not really translatable" (Bienek 1985, p. 7) or even untranslatable. It is true: the word is difficult to translate; in its condensed fusion of meanings created by poetry, it is probably actually a fairly German affair. This is how Rolf Petri walks through the difficulties of the translation: the English *homeland* is as emotionally positive as *Heimat* and close to the substantial exchange value of the word in the nineteenth century; the French terms *le petit patrie* (small fatherland) and *matrie* (motherland) come close to some of what is meant by *Heimat*, while "*le pays*" (the land) expresses the appreciation of the regional; in Italian, *paese, nazione, paese natio, terra natia,* or *patria* could be aids—but just aids. Some of the translations contain less, some more, and some other meanings than *Heimat*. From this, Petri concludes that "the serious translation problems are based on a historically relevant differentiation of the terms" (Petri 2001, pp. 80 f.). He is referring to Hermann Bausinger and his attempt to demythologize the untranslatability of *Heimat*: It was a matter of ordinary translation difficulties, as with many other words.

The dilemma resolves by taking apart meaning and significance. Petri admits, "It would be nonsensical to view the need for manageability and participation in the spatial articulation of social relationships as a 'German' need" (ibid). Bausinger had meant nothing else: "The thesis of the untranslatability of *Heimat* comes down to the assumption that people elsewhere did not develop a particularly intimate relationship with the place where they grew up or where they live.

This is certainly not the case, even if the bond is not equally strong and durable everywhere. What is difficult to translate from the term *Heimat* is the general feelings of personal appropriation of a place or a landscape. Rather, it is the specific coloring of these feelings; it is the romantic mortgages of the *Heimat* term that play a special role in Germany." (Bausinger 2000, p. 72).

A Seducative Concept for Ethnic Minorities

The idea of *Heimat* is particularly dense in South Tyrol/Italy. It is considered God-given in a novel and film title by the alpine-hero Luis Trenker (*"Heimat* from God's Hand*,"* dedicated to his mother, cf. Trenker 1979, p. 448). In its political meanings, *Heimat* invokes the (supposedly) good old days ("South Tyrol – German for 2000 years" as a car sticker), becoming a kind of political sacrificial altar in front of which new generations are kneeling. "No victim too difficult for *Heimat* ..." is the title of a documentary about the imprisoned and partly tortured bombers against Italian politics in the 1960s (Golowitsch 2009). "Friend, you who still sees the sun, greet me the *Heimat* that I loved more than my life," the combatant Luis Amplatz chose as the theme for his grave (Peterlini H.K. 2021a, p. 356). For the South Tyrolean riflemen, who see themselves as the bearers of the Tyrolean tradition of fighting for freedom against Napoleon in 1809, *Heimat* is a political creed that culminates in the militant circles of the riflemen movement in the separatist demand for self-determination for South Tyrol regarding its future state affiliation. The riflemen see themselves based on tradition and keeping their self-image alive as carriers of a cross-generational mission and legacy to protect their *Heimat*.

From the statute of the South Tyrolean Riflemen Association[3]:

> *"The purpose of the federal government and the riflemen companies and riflemen chapels connected to it is:*

---

[3] The statute reproduced here is the one that was valid for this work during the survey period; a new version was approved at the Federal Assembly on April 27, 2019, in Bolzano; the changes to the passage quoted are set out in the footnotes below; the change of the point of maintenance to "the native natural and cultural landscape" (from the last to the second line) made it possible to point to the maintenance of "target shooting" (as a traditional shooting sport activity) together with customs and costumes.

– *Faithfulness to God, adherence to the Christian faith – traditional beliefs – and the spiritual and cultural heritage of the ancestors; Protection of the Heimat and the Tyrolean way of life and nature*[4]*;*
– *The unity of the state of Tyrol, the exemplary exercise of the rights and obligations of the South Tyroleans to preserve the Tyrolean character and to secure the livelihood of the German and Ladin ethnic groups in their ancestral Heimat;*
– *Human freedom and dignity;*
– The care of the Tyrolean shooting rituals, the traditional costumes, and the native landscape and nature.[5]

Maintaining these principles is the highest obligation of Tyrolean riflemen." (Schützenbund 2020)

In an introductory preamble on the website of the South Tyrolean Riflemen Association, it says:

"The task of the riflemen today is to defend Tyrolean identity whenever it is threatened. Identity is defined as the language, culture, custom, legal sense, belief, value system, and generally practiced norms of people's behavior in a certain area. Identity is the sum of the characteristics passed down through generations that have shaped people in a certain area (Heimat) and give them an unmistakable face. The geographical area of these people – their Heimat – is shaped on the one hand by these people; on the other hand, Heimat shapes the people. From this perspective, riflemen are active Heimat guards!"[6]

Sketches of a Longing: Psychoanalysis as a Model of Understanding

Psychoanalysis offers a wide range of instruments for understanding the creation of national identities, such as that associated with *Heimat*, especially in its cultural-theoretical and ethno-psychoanalytical branches (cf. Peterlini H.K. 2010b). Sigmund Freud has already created the basis for researching cultural, political, and social phenomena through psychoanalytic empathy, interpretation, and understanding; the personal psychological, and collective formation of myths was a matter of concern and source of knowledge in many of his works.

---

[4] In 2019, supplemented by "and the native natural and cultural landscape.".
[5] "And the native landscape and nature" in 2019 was replaced by "as well as target shooting.".
[6] This passage remained unchanged despite a modernization of the website.

A prerequisite for the formation of enemy images and phantasms are psychic processes of repression, splitting off, and projection of burdensome parts of the ego (Freud 1911/1958). Ultimately, it is the ability and disposition of human beings to displace, detach from, and project to the outside what is incriminating, ineffable, inexplicable, frightening, and guilty. Freud already gradually relativized his original trauma theory whereby only severe, sexual traumatization of the infant triggers this mechanism. After him, generations of psychoanalysts (Ferenczi 1949; Klein 1946; Klein and Riviere 1953; Fairbairn 1952; Winnicott 1965; Richter 1972; Kernberg 1975; Stern 1977) developed the model in different directions. Not only the violent assault (this especially, of course) but also general existential experiences would be difficult to cope with for individuals and groups if they. National and ethnic conflicts are—as Josef Berghold (2005, pp. 110 f.) shows in nine theses—an ideal projection surface for psychological stresses that are difficult to manage privately.

One of the central psychoanalytic explanatory models for the projection process goes back to Freud's student Melanie Klein. In its intensive interaction with the mother's breast, the infant experiences the mother's breast as present and caring (and therefore as a good breast), absent and failing (and thus as a bad breast). According to Melanie Klein, the ability of "projective identification" develops based on this experience of care and refusal by one and the same object, which the child still experiences as part of himself (Klein 1946, p. 102). For the infant who has no or minimal rational explanatory patterns, the experience of the absent breast goes along with the deepest fears of exposure, hunger, and death, which the baby cannot process rationally in this phase of development. Klein observed two successive phases in the infant's reactions, the *depressive* and the *paranoid-schizoid position*. In the depressed position, the absent or failing breast is encountered in the hope of reconciliation; aggression and accusations of guilt are not directed outwards but swallowed (*introjected*). In the paranoid-schizoid position, on the other hand, with help from the projection, "all painful and unpleasant sensations or feelings in the mind are by this device automatically relegated outside oneself; one assumes that they belong elsewhere, not in oneself. We disown and repudiate them as emanating from ourselves; in the ungrammatical but psychologically accurate phrase, we blame them on to someone else." (Klein and Riviere 1953, p. 11). Since the child would not be able to cope with the ambivalent (ambiguous) sensation compared to a once good, sometimes bad mother's breast, a "bad object" (ibid, p. 22) is selected onto which the aggression can be projected. For Klein, the healthy development went from a paranoid-schizoid position to a depressive one, in practical terms, from acting out anger towards others toward taking responsibility for oneself.

In interpreting and further developing Klein's model, Wilfried R. Bion understood the continuing oscillation between depressive and paranoid-schizoid positions as a lifelong cycle of development, which occurs as a dynamic process between disintegration and integration, fragmentation and coherence, chaos and formation. Mental illness, therefore, entails "to make a dynamic situation static," while recovery entails "to restore dynamic to a static situation and so make development possible" (Bion 1963, p. 60). Accordingly, the flexible, agile change from a depressive to a paranoid-schizoid state and vice versa is not dangerous; it ultimately represents psychological agility and enables handling of the environment, fellow human beings, and life situations (cf. Lahme-Gronostaj 2003, p. 66). The life-aggravating problem arises only from an increase in one or the other position in the extreme and a fixation on this state. From an individual-psychological point of view, this would be the suffering of depression or psychosis, i.e., in self-paralysis through guilt and shame or in the delusional accusation of external objects that supposedly (or in part also in reality) are endangering oneself.

From Inner Soul Questions to Social and Political Patterns

What is significant about the psychoanalytic models is that the toddlers internalize their misery and their reactions so profoundly due to the not yet possible reflexive processing. Consequently, these experiences inscribe themselves as patterns in body and soul (considered as unity in phenomenology). With similarly triggering moments and missing favorable resolutions through life and learning experiences, these can reactivate themselves at any time and reproduce for life.

For social processes, we could deduce an exciting hypothesis. An open discourse, which enables and encourages open and democratic reflections on the crisis, prevents the emotional depressive congestion that would otherwise result in angry, psychotic reactions. If discomfort can be articulated and communicated, if different positions can be negotiated, and angry feelings have their place, it may be possible that the valuable parts of the depressive phase—acceptance of guilt, responsibility, willingness to compromise—come into play. One could object that the currently free discourse in digital media makes this participation possible. Therefore, these new possibilities for almost all people should lead to an acceptance of foreignness. Why then are straightforward xenophobia and populist agitation raging in social media, as discussed in the chapter "Dialogue with Adorno" in this volume? At this point, it should only be noted that the discourses in digital media rarely result in an interpersonal and reflective, nor self-reflective exchange, but mostly get caught up in irreconcilable pro and counter positions, so communication lacks a responsive character and an empathetic understanding. The hate messages on social media are not a reflexive treatment of fears and their examination from the perspective of life questions and life situations, but rather the angry crying of the baby, who cannot explain his misery, impotence, and helplessness. It is difficult to predict if we learn to use the new media in a more constructive way for personal exchange and social reflexivity instead of aggression.

A historical example could give hope: In its first application, the radio was also a medium for the propaganda stimulation of the masses by fascism and Nazism until it was used by all population groups and used for its potential to promote democracy. The stimulating, more shouted than spoken, speeches by Hitler and Goebbels on the radio have given way, albeit dramatically late, to interviews with people from different backgrounds and positions, reports on life stories, living environments, and specific problems and possible solutions.

Therefore, the decisive factor would be whether people with their actual questions, concerns, and issues are heard and recognized in social exchange (cf. Honneth 1995), avoiding the shift to irrational fears and enemy-images of the Other, while actual problems remain unspoken and unresolved. For social work in the migration-shaped society, racism has become a new challenge (cf. Geisen 2018). Justified fears about the future (i.e., due to the no longer undoubtedly secure prosperity in Europe, the limits and consequences of a capitalist economy built on growth and the planet's future) are often projected onto migrants and refugees. According to the psychoanalytical theory about such dynamics, the so constructed enemies of *Heimat* are not triggers, but at most, symptom carriers of deeper and repressed problems.

The female connoted Heimat as male matter
In a small, manageable land like South Tyrol, these dynamics can be studied almost in a test tube, including their historical dimension, which affects the present day. The remembered history of Tyrol, which is very present as a myth and which reinforces defensive attitudes by reactivating old patterns, is a narrative of suffered oppressions, followed by liberating outbreaks. At the same time, the inner discourse in Tyrolean/South Tyrolean society was—and is—affected by an identity model that depicts identity primarily as a unity, first in a religious, later in a political and linguistic, and finally in an ethnic sense. Thus the political myth of Tyrol is a myth with high unity pressure, paralyzing the communication within the group, with the danger of long-lasting resentment and eruptive discharges.

A key topic for *Heimat* psychoanalysis is birth trauma (Rank 1924/1993) as a formative experience of loss of security, which constantly reactivates itself in the course of life. Leaving a familiar environment is, on the one hand, a prerequisite for developing and living; on the other hand, people experience it as a loss of care, support, and security (however deceptive these may have been). The birth is highly distressing for the fetus and an existential threat that simultaneously brings him to life. The parturition does not necessarily have to be extraordinarily traumatic. Even under reasonably normal conditions, for the fetus, it represents the leaving of comforting care and places the fetus in a completely new situation. *Heimat* is a metaphor for both: as a womb symbol, it is the great mother, who protects from exposure to life and yet has to be left again and again for an independent life; as the last or eternal *Heimat*, it promises a return to the lost paradise at the price of life.

Especially in the riflemen tradition (which is ultimately the most direct expression of the Tyrolean defense culture), *Heimat*—in German a feminine—mostly appears as a maternal myth. Still, it is a purely male affair in its political design and military defense. Ultimately, *Heimat* was "owned by the powerful men" (Unterrichter 2007, p. 13). In 2006, for example, there was still an intense debate in the South Tyrolean Riflemen Association about whether women should be more than just marketers among riflemen. Women in the company once had to assist the soldiers in the war with consolation, bandages, and gifts of love. More recently, they have been marching along as adornments and pourers of grog (cf. *Dolomiten* 2006a, p. 15, b, p. 15). The equal opportunities councilor, Julia Unterberger, urged the riflemen to abandon a gender role tradition according to which their marketers have historically been "for the most part prostitutes who went with men to war." (*NTZ* 2006a) The Minister of Culture, Sabina Kasslatter Mur, asked the riflemen association to "rethink its role assignments, which may have been justified centuries ago, in the spirit of current developments" (*NTZ* 2006b).

The dispute initially flattened to no avail; women's equality in the riflemen organization continued to fail. However, the marketers organized themselves at a first country meeting on August 1, 2015. The resolution passed at the time states: "We marketers no longer see it as our sole task to be an eye-catcher in the front row next to the captain. We need to contribute to the community's success by actively participating, contributing our ideas, and taking responsibility. [...] We marketers know that male comrades expect a lot from us: correct, clean appearance, punctuality and reliability, a sense of duty, and comradeship. We also want to make a difference for *Heimat* and culture. We take on various tasks such as maintaining the traditional costumes, shrubs, and crosses. We marketers march in the front row next to the captain; we are very well aware of this and proud of it. We are proud of our beautiful costumes." (Schützenbund 2019).

As a result, the marketers were given their own statutes (ibid) and their own regulations. It says, among other things: "The costume should be freshly ironed if possible. The tips of the blouse should be strengthened. The costume itself is jewelry, so pay attention to your jewelry; it should be subtle and match the costume. Particularly striking and prominent colors and shapes, oversized wristwatches, and many rings are unsuitable and should be avoided. The costume needs matching shoes (costume shoes). Fashion shoes with high heels should be avoided. The hair should always be neatly styled and, if it reaches over your shoulder, tied up, braided or tied together (also under your hat) so that you can see the beautiful costume, but above all, your face. The make-up should be applied discreetly. Particularly noticeable red fingernails or red lips are not part of the costume and are inappropriate." (Ibid) No dress and outfit code could be found on the same website for male shooters.

Narcissistic theory as model of understanding for the shift from social fears to national unity
Heimat can also be understood psychoanalytically as a narcissistic draft of a dream world in which the individual and the group know that they are safe and loved. Negative and narcissistic projection (cf. Volkan and Ast 1994) complement each other: everything that is good for the self promotes the idea of an intact identity. What has to be pushed out of the actual experience is projected onto the fear and enemy images, the phantasms. As uncertain as the emergence of the term *Heimat* may ultimately be (cf. Bausinger 1980, p. 16), its fascination is undoubtedly unfolded by overlapping breaks and losses. The vision of deep security and solidarity emerged from a set of regulations, the *Heimat* law, which granted the right of establishment and purchase to one, and denied it to the other. The questioning of the established world view of a vault between heaven and hell through the Enlightenment, the alienation of people and living space through industrialization and technology was contrasted with the *Heimat* idea as a "processing form of the experiences of loss" (cf. Heinz et al. 1980, p. 45).

In the exaggeration of the nation towards the end of the nineteenth century, *Heimat* became an emotional and political shelter against the other. According to Mario Erdheim, the national idea cannot be viewed separately from suppressing social fractures and conflicts. In its psychological need for protection, the individual avoids everything that could bring conflicts with the group and its ruling authority—also in the interest of the applicable rule and economic system (Erdheim 1984, p. 25). Breaking out of the unity (identity) of the group harbors the risk of social death, which, depending on time and political system, could also cost life in real terms. The perception of social disparities within the group threatens their unity, while the shift of discontent to national issues and external enemies strengthens internal cohesion.

The need for groups to cover up internal breaks in favor of an illusionary homogeneity also explains, as Alexander Mitscherlich argues, that the fears and hostility that are repressed as a result are projected onto scapegoats and strangers outside the community (Mitscherlich 1969, p. 31). For example, national conflicts are also staged unconsciously to act out individual psychological conflicts that are reflexively inaccessible to the individual. The national agitation in the collective relieves the individual of a reflexive examination of their own problems, life situations, and worries.

The feeling of national solidarity compensates the individual for accepted social disadvantage, incapacitation, and exploitation. Considering that the Latin term for birth (*natio*) is included in nation, it becomes plausible that the nation is a fitting mother projection, an idea of collective security, of being collectively cared for, and cared for by a big, good mother (cf. Aigner 2002, pp. 253 f.). Everything that threatens this idea, which life inevitably brings with it, such as failures, insults, and social fears, is split off by the good mother—and projected onto an external enemy (ibid, p. 303). In the social dimension, the actual Other is only perceived in his—oversized—menace for the nation as a symbolic mother.

For the Tyroleans, who always wanted to be their own *nation*, the equivalent concept for the nation is *Heimat*—with all its translational difficulties. What specifically means the home courtyard as neuter in the dialect (the *Hoamat*), as a feminine (*the Heimat*) is a territorially, culturally, politically, and nationally expanded idea of security. *Heimat* is the condensation of all positive national feelings, a good mother who creates security, is lovable and pure, and must be protected from foreign intruders/rapists. The national struggle for this *Heimat* makes it possible to channel all feelings that do not fit in, such as suppressed fears, frustrations, feelings of guilt, into aggression against foreign enemy images, to remain in tune with one's system of rule and to remain united as a population group. The identity of the individual, which is itself a psychic construct, is partly founded, partly strengthened, partially armored, and hardened by the collective identity.

From my family history, two episodes of talks with my mother come to my mind, demonstrating how offenses can be shifted in perception and memory. I had always suspected that my mother's regret for missing educational opportunities ("I would have loved to study") was due to fascist oppression. The ban on the German school forced her to go to the Italian school. This may have played a part without a doubt, but on the other hand, my mother made good progress in the Italian school. The main reason for not having gone to secondary school was that an aunt of my mother persuaded her parents that she had a suitable apprenticeship with a German-speaking merchant family at the time, who "then took serious advantage of me."[7] The parents, who at the time could have afforded secondary school because of their thrift, had made another good-faith decision that deprived my mother of all of her educational aspirations as an apprentice; it was not Italian fascism, which was undoubtedly an oppressive system, but that the country's own people were significantly involved in this through an exploitative attitude towards female apprentices. This exploitation was more challenging to admit than the (justified) denunciation of fascism, an externalized enemy. The second example: In the interwar period, my grandfather did well professionally due to his manual skills. His specialty as a carpenter was building stairs. However, he lost a good job at a carpentry workshop because he could not cope with a foreperson. When he wrote a somewhat undiplomatic letter to company management, he was released. My mother only told me both episodes in old age, and even though she may have described me earlier, I had not fully assessed it but rather related both injustices exclusively to fascism. The image of a fascist foreign power that suppressed my family was too strong to recognize the proportions of disadvantage and injustice that should be assigned to our system. It may also be easier to bear insults and the shame associated with them if they are pushed towards the foreign power and shared collectively and in solidarity with your own group. This, in turn, can explain why political debates, in particular, are a projection screen for the disposal of psychological burdens—individuals see their suffering socially divided in the struggle with their group against another. If the injustice of one's own group was recognized, the individual would have to deal with it and risk social exclusion.

Enemy images between reality and imagination
The hypothesis that ethnic groups (cf. Volkan 2003) can go through similar psychological phases like individuals is a premise for psycho-historical and psychoanalytic interpretations of social developments. This assumption does not deny the actual historical events but perceives them as more complex emotional and psychological charges. Freud used the concept of *over-determination* in the sense of multiple determinations and ambiguity for the coexistence and overlapping of equal causes to create the unconscious dynamics (dream structures, symptoms, repressions, separations, distortions, displacements, projections) (Freud 1900/1913, 1915/75). Accordingly, historical facts cannot be analyzed independently but are always contingent on their interpretation. According to Derrida, the thing itself is always to be understood with meanings: "There is

---

[7] Personal minutes of the interview, March 2010.

no absolute origin of sense in general" (Derrida 1997, p. 66). Only traces can be questioned—as Derrida alludes to the phenomenology of Husserl—for their possible meanings, namely in the sense that "[t]he trace is the différance which opens appearance (l'apparaître) and signification" (ibid).

Fact and interpretation, event and meaning, appearance, and signification cannot be separated from one another even by the most careful historical research. From a psychoanalytic point of view, creating enemy images is easier the more they combine with actual events. Immigrants from Italian provinces in the 1950s were, in the popular perception of South Tyrol, representatives of the ruling, patronizing state. It was easier to recognize and act out against strangers, who were also state-skilled and privileged, what had to be suppressed in one's own: the social competition in one's group, the loss of importance of the rural economic structure with existential insecurities and insults in the middle of the last century, the patronizing and exploiting by South Tyrolean employers and/or by an oppressive father, for whom the foreign state was a grateful replacement object. In some biographies of those assassins who fought for South Tyrol's freedom with bombs in the 1960s, there are depictions of conflicts with their own fathers or the shame about their failures (cf. Peterlini H.K. 2010b, pp. 60 ff.). The fact that the state acted in a truly oppressive and authoritarian manner facilitated the shifting of aggression and made it a grateful projection screen for what had to be swallowed about one's own father and—due to the child's love for the father—was not accessible reflexively.

Similar observations can be made in the current migration debate. Concerning migrants, fears of economic overreaching are particularly prominent in the factual argumentation. These may be partially and occasionally justified due to the higher willingness of foreigners to undertake work that requires lower wages or the higher demands on social benefits due to greater need (Berghold 2005, p. 153). However, the psychotic distortion of real people to the phantasm "foreigners" makes it almost impossible to see them not only as competitors or intruders but also as service providers, taxpayers, fellow citizens, and people. The foreigner serves as a lightning rod for what is scary and troublesome.

Cultural memory and its interacting with collective traumata

An essential link between the individual's psychological processes and the large group is memory culture. Whether and how something is remembered and what, on the other hand, falls into oblivion, is fundamental to collective identities. Thereby, the memory and repression performance of the individual interacts with that of the group. According to Volkan, the bond between the individual and the group is woven primarily through inner images of the history of large groups such as myths, songs, eating habits, dances, heroes, and martyrs (cf. Volkan 2003). This is where the psychoanalytic theory of forgetting (through splitting off and replacement) meets Jan Assmann's theory of memory. In continuing the idea of collective memory by Maurice Halbwachs (1980), Assmann distinguishes four external dimensions of memory:

- Mimetic memory—this refers to action and intends different forms of behavior through imitation; we found it saved in instructions relating to machinery, cooking, construction;
- The memory of things: from private everyday objects such as beds, chairs, crockery, clothes, and tools to houses, streets, villages, towns, cars, and ships;
- Communicative memory: Language and the ability to communicate with others;
- Cultural memory—the handing down of meaning. (Assmann J. 2011, pp. 5–6)

Like every subdivision, it is prone to gradual or abrupt transitions. In this way, areas of the "mimetic memory" can become routine, become a rite, thereby exceeding the pure memory of activity and gaining a special meaning. As an example, it occurs to me that when my father cooked the polenta, typical corn-meal from Trentino and the South Tyrolean lowlands, my father always ended up drawing a cross on the simmering polenta with the salt and often explained to me that this was what Nonna did, his mother. The salting of the polenta had become, through the adoption of a sign, a ritual which—from Nonna to my father, from my father to me, from me to my children—was adopted into cultural memory and assumed an emotionally charged meaning that exceeded the pure function of salting.

A collective does not have a substantial memory of its own, but "it determines the memory of its members" (ibid, p. 22). The individual memory is also socially constituted, as it includes what others say, how they say it to you, and how they mirror what you have said. Assmann distinguishes communicative collective memory from collective cultural memory. Where the former includes those mem-ories that contemporary witnesses share and are therefore no more than 80 years old, cultural memory "focuses on fixed points in the past" (ibid, p. 37), like "tales of patriarchs, the exodus, wandering in the desert, conquest" (ibid). In cultural memory, factual history is transformed into remembered history and then myths. Assmann thus outlines the importance of the culturally remembered past in myth for the present: only through (selective) remembering does history become a myth, and only as a myth does the past regain meaning for the present (ibid, pp. 37–38).

The ritualization serves the role of memory support. While the communicative memory flows in everyday life, the cultural memory is anchored in fixed forms. It has no neural basis but is carried by myths, rites, history books, teaching, films, media, monuments, the narratives of architecture, and all other fixed forms of tradition. Canonical and barely changeable texts such as the Bible and the Koran, canonized historiography, practices, and customs form the collective consciousness.

Rites and festivals are of fundamental importance in the Tyrolean memory, which is cultivated particularly intensively by the riflemen. Military readiness was provoked by prohibitions or restrictions of rites, very close to Melanie Klein's concept. It is mostly an external attack that causes a child to change from a depressed to a paranoid-schizoid mood. The uprising of 1809 against the Napoleon-Bavarian foreign occupation was triggered mainly by attacks on rites and cultural gatherings, the ban on the midnight mass, and the transfer of the Tyrolean to the Bavarian army. The military mustering shows the state's paternal power, which judges whether the young man's testicles are suitable for military service—the defense of the mother's home. In the village of Tramin, where customs have a robust patriotic note, young men painted a Tyrolean eagle on their doorstep before the mustering for the determination of military fitness (until mandatory military service in Italy was abolished in 2005). It's easy to interpret it as the symbolization of a sense of *Heimat* against the compulsory service with the foreign state (Peterlini H.K. 2000). The most excruciating memories from fascism have to do with the prohibition of German songs, the translation of German epitaphs into Italian, the tearing down of the typical red and white buttons from the costume (Peterlini H.K. 2021a, pp. 17 f.). The first bomb attack in the 1950s was grounded in the fact that a traditional dance during the annual Bozen trade fair procession should no longer be performed by South Tyrolean guys but by Italian recruits (ibid, pp. 28 ff.). In this way, the familiar South Tyrolean community's economic hardships and social breaks were redirected to a state that offered itself as an enemy due to its lack of sensitivity and toughness towards its minorities.

Motives of impotence, oppression, and defeat, as described by Wolfgang Schivelbusch (2003) in "The Cultures of Defeat," play an essential role in this regard. Societies whose collective memory sticks to repressed frustrating situations of helplessness and defeat in the past are particularly resistant to change. They hold on to their comforting and glorifying myths to an increasing extent since the pain of defeat cannot be released for conscious processing. A strategy to alleviate defeat—closely intertwined with the myth of the hero—is creating the mythical figure of treason. The betrayal clears the defeat of one's fault and blames a traitor's failure. Tyrol's social history, which has been well revised but hidden from myth, shows patterns of the mythical primary figure of the sacrificed, abandoned, betrayed hero. The rural family structure in Tyrol meant that out of the necessity for survival, only one—for a long time the youngest, later the oldest of the mainly many children—could take over the *Hoamat* (the farm). The others were forced to decline socially. If they were fortunate, they could learn a trade or were married well if they were daughters; in the less good and more frequent case, they became bondsmen and bondswomen on the farm of their lucky brother or in the yard of a neighbor. They had lost *Hoamat*, their home-world, because of a father, who was usually authoritarian but was economically too weak to protect all of his children. In areas with real division, the courtyards were so fragmented that no one could live on them. Thus, the Upper Inn Valley, Ausserfern, Vinschgau Valley (Venosta Valley), and the Italian-speaking part of Tyrol (Welschtirol[8] or Trentino) became regions of crisis and emigration.

The myth of the Tyrolean national defense has also been heroically charged with devastating and, as a result, mostly extremely socially stressful defeats. The myth is said to at least alleviate the severity of the defeat. The loss of defeated countries, states, cultures reflects the fear of defeat in general—not only of the people but also of the individual: in the social struggle for survival, in the vulnerability of life to illness and death. The defeat would require an acceptance of guilt, shame, fate (in the sense of what happened), and mortality, but is instead suppressed, so their refinement gives strength for life. Martyrdom is one way of dealing with the sure defeat inherent in human mortality: Christ is dead for three days until the resurrection rebuilds the world. Andreas Hofer became the hero of Tyrol through execution (cf. Larcher 2005, pp. 182 ff.).

The Hofer myth, for example, shows Freud's interpretation of the myth as *poetic fancies* of the past into grand narratives with which communities adjust their history. They comfort over hurtful, distressing, troubling, unresolved, guilty elements in their history but do not heal them because they suppress painful or regretful parts and elude conscious processing. In the Hofer myth, the Tyroleans soothe—from a psycho-historical point of view—their history as residents of a small and always occupied passport country. From a social-psychological point of view, they alleviate the social misery that in 1809 depressed the country far more than the culturally and linguistically familiar Bavarian occupation. And they

---

[8] Welschtirol was the definition for the Italian part of the "Land Tirol" during the Austro-Hungarian-Monarchy; concerning this 'old Tyrol,' the term is still in use, officially the area is now called Trentino or Province of Trento according to its central city Trento.

appease—from an individual psychological point of view—their fear of vulnerability and death. The myth of a *Heimat* that needs to be defended against foreign influences and intruders, however, also reduced the ability to integrate "good" and "bad" parts, prevented (and still essentially prevents) an internal debate about individual freedoms and social rights in Tyrol.

This is the tricky thing with *the manifestation of distorted cures* by splitting off, repressing, and projecting: It does relieve you of inner pressure, fears, feelings of guilt, and aggression. But by disposing of all justified dissatisfaction and all internal contradictions on the intruder from the outside, outdated patterns are solidified, the concrete examination of one's reality withers. Since this process is largely unconscious, it is beyond reflection, communication and thus also beyond active intervention.

The ambivalence between relief and constraint from identity offers such as Heimat
Heimat as a symbolic idea of an ideal world, for which one goes to war or surrenders faithfully as a wife and mother, was one way of coping with actual *Heimat* losses. At the same time, it could have been a relief to project the aggression that had to be suppressed against one's own—ultimately innocent—father, against one's own economic and rule system, onto external enemies (Peterlini H.K. 2010b, pp. 75 ff.). It is easier to live when your imagination—concentrated in *Heimat*—is intact and the misfortune is caused by a foreign power, by a fate that is not well-meaning. Efforts for political change are more likely to be avoided for the sake of harmony with one's group unless the danger comes from the outside. The closeness of the personal live-experiences for generations and the political victim history of the country is illustrated in a metaphor by the South Tyrolean historian Claus Gatterer about the South Tyrolean lifestyle after the annexation: "The path from Tyrolean to South Tyrolean was one descent, a downgrading. It was the way from the lord to the bondsman." (Gatterer 2003, p. 15). A descent to the farmhand (and to the maidservant) was also a very individual fate for most farmers' children.

Myths also offer models of action and roles, which are easier to slip into—when the old patterns are reactivated—than new solutions that would first have to be tried out. The South Tyrolean dynamite attacks of the 1960s (cf. Peterlini H.K. 2021a) can be seen as the latest violent reactivation of the Tyrolean defense myth. In the guerrilla-like uprising of 1809, the attackers in the 1960s found 150 years later a mythically transfigured pattern from their history that they could use. What used to be the pitchforks and stone avalanches in 1809 were primarily self-made explosive sticks in the 1950s. "There was no other way" is the title of a book published by the Association of Former Political Prisoners (Mitterhofer and Obwegs

2000). The myth didn't let them see any other way. Role models and myths were missing for different strategies, such as passive resistance or civil society protest; they only knew (and know) from their history, long-endurance, and sudden outbreaks with mostly helpless but uninhibited violence. A later series of attacks in the 1980s with iron tube bombs and machine guns—partly stimulated by provocateurs—shows how such role models do not lose their fascination even if they seem to be out of date.

Some young people got inspired, some old bombers wanted to go for it (Peterlini H.K. 1992, pp. 51–73) though South Tyrol in the 1980s had a solid political Autonomy status (cf. Peterlini O. 2009). The Tyrolean leader against Napoleon Andreas Hofer is still a strong identification figure among young shooters and marketers in the twenty-first century. The state is seen as the enemy, an intruder into the *Heimat*, frightening and challenging. At the same time, the state is also a possibility for a contra-phobic attitude, as shown by the statement "Südtirol ist nicht Italien" ("South Tyrol is not Italy") provocatively placed on the Austrian-Italian border on the Brenner.

Such a country offers itself as a laboratory for the attempt to understand the ambivalences of *Heimat*. The theoretical approaches will be reflected in their relevance for political and personal identity formation in the following chapter.

Who Am I? And Who Are You? *Identity as a Construct Between Self-Invention and Straitjacket*

The Presumed Unity of Person and Society—and its Fragility in Theory and Real Life

"We do not want to live in a mishmash, we see everywhere what that brings. In a purely German valley, there will be no problems." (Peterlini H.K. 1998, p. 83) This statement by a South Tyrolean young rifleman at an educational conference of the South Tyrolean Riflemen Association on February 14th and 15th, 1998 in Brixen stems from a term which riflemen pronounce almost as solemnly as *Heimat*: identity.

From a keynote speech by former state commander Richard Piock:

> *"It is the many small duties and activities of the shooters who, in a time of escape from responsibility and selfish liberalism, have been given that unmistakable face of the country that is commonly called identity: in many cases, processions would no longer be held, costumes would have disappeared from the image of Tyrolean society, war cemeteries would not be maintained, history would not be remembered, the fallen would no longer be remembered; it is the statements of the riflemen in public about wrong decisions made by politicians and indifference to society that are supposed to prevent us from losing this identity; identity is not a postcard idyll from enterprising tourism prostitution – identity is a distinctive way of life and attitude to life, based on tradition; Identity is not just language, custom, and customs, but also basic ethical behavior such as helping neighbors and solidarity. Preserving identity also means not selling out Heimat, not sprawling, not concreting it, and not exploiting it. Riflemen have always had this function of identity protection – 1703 or 1796, 1799, 1809 or in the period from 1914 to 1918."*[1]

---

[1] It was the speech at the Federal General Assembly in April 1999; Copy owned by the author.

© The Author(s) 2023
H. K. Peterlini, *Learning Diversity*,
https://doi.org/10.1007/978-3-658-40548-9_3

As a state commander, Richard Piock was part of a tradition of quite eloquent and intellectually shrewd minds, like before him the politicians Karl Mitterdorfer as South Tyrolean Senator in the Italian Parliament and Bruno Hosp, the long-time state secretary of the South Tyrolean majority party SVP (South Tyrolean People's Party). Piock was always culturally open-minded as a rifle officer and encouraged cross-spirits and critical contributions to Tyrol. As general manager of the industrial company Durst Phototechnik, which operates internationally from Brixen (Italy) and later also from Lienz (Austria), he moved globally and interculturally; as governor of the riflemen, he stood for what was local and monocultural: "In a time of anonymity and increasing decay of values, we have an increasingly big task to fulfill – we have to make sure that the young people find the way to Tyrolean identity and not lose themselves in a multicultural world without value and meaning."

Identity thus appears as a condensed structure of values and meaning from customs and tradition, but also nature, landscape, ethical values, delimited by a "multicultural world without value and meaning," marked with the years of wars against aggressors from outside: 1703 against Bavaria, 1796–1799–1809 against Bavaria and French, in the First World War against Italy.

"The ego can identify with itself precisely in distinguishing the merely subjective from the nonsubjective," Habermas summarizes the theories "from Hegel through Freud to Piaget, that subject and object are reciprocally constituted, that the subject can grasp hold of itself only in relation to and by way of the construction of an objective world" (Habermas 1979b, p. 100). For Habermas, essential functions of identity formation are the creation of a psychological and social unity through fixed worldviews, which were of "original actions of myth having been transformed into 'beginnings' of argumentation, beyond which one cannot go" (ibid, p. 104): "The unifying power of worldviews is directed not only against cognitive dissonance but also against social disintegration." (ibid, pp. 102–103) Ultimately, it is about "the unity of the world," whereby "homologies" could be adopted between the ego and worldview structures (ibid, pp. 105 f.). By comparing two statements, Habermas shows how collective identity, in contrast to individual identity, is no longer dependent on dialogue with a You (in the singular and plural):

Sentence 1: We participated in the demonstration (while you stayed at home).
Sentence 2: We are all in the same boat.

While the first sentence aims at a 'You' group outside of the '*We*,' the second sentence only addresses the 'We' members. For Habermas, such sentences have the meaning of self-identification: "We are X (where X can signify Germans, citizens of Hamburg, women, redheads, workers, blacks and so on)." (Ibid, p. 108) The individual ego needs the affirmation of a '*You*,' while the '*We*' is self-sufficed since it receives the approval from within from the members. In relation to *You* (in the plural), the relationships, thought abstractly, can only be limited to delimitation: "A group can understand and define itself so exclusively as a totality that they live in the idea of embracing all possible participants in interaction, whereas every thing that doesn't belong thereto becomes a neuter, about which one can make statements in the third person, but with which one cannot take up inter-personal relations in the strict sense – as was the case, for instance, with the barbarians on the borders of the ancient civilizations." (Ibid).

Habermas steps through the identity models of psychology, psychoanalysis, and sociology by playing through possibilities of how identity arises and acquires importance. He agrees with George H. Mead that the "unity of the person" is built through "inter-subjectively recognized self-identification," through belonging to and delimitating from a group. Mead had assumed the "I" as a personal, pre-social, and subjective instance of the personality, while the "Me" was the side of the personality shaped or mirrored by society. Mead also used psychological models: "If we use a Freudian expression, the 'me' is in a certain sense a censor. [...] Social control is the expression of the 'me' over against the expression of the 'I.'" (Mead G. 1934, p. 210). Identity as a unit of the self constantly forms in chain processes a new balance between the two ego instances, whereby the social character of the individual is decisive. In contrast, the individual ego creates individuality through its impulsive reactions to society.

The fragile construction of presumed strong identity
Thus, albeit modified and transferred to other models, Habermas and Mead refer to psychoanalytic identity models. The assumption of a subject that is not sole and holistic but instead divided and branching into often broken dimensions unites almost all psychoanalytic schools, no matter their differences. We find them in Freud's structural theory of the psyche with its three instances id, ego, superego, which are partly hostile to each other (cf. Freud 1923/1975). We find them in the basic assumption of the unconscious, which, according to Freud, can never be entirely brought into the conscious and which makes people feel like strangers to themselves (cf. Freud 1914/1975). We find them in the Freudian sexual theory with its partial drives, which are only painstakingly put together to create more or less fulfilling and socially acceptable sexuality (cf. Freud 1905/1962). And we find them in Freud's theory of culture, which assumes that libidinal and aggressive drivers are difficult to keep under a socially woven blanket of culture; even briefly

venting this civilizing blanket "reveals man as a savage beast" (Freud 1930/1961, p. 58). Accordingly, we would not be what we think to be. Still, we must find contradictory subjects, who are constantly redefining and reorienting themselves, who have to construct their identity (their oneness with themselves and others) with difficulty and are therefore so anxious and clinging to it.

The great myths also talk about it: In Biblical Genesis, it is the expulsion of man from paradise that gives him the freedom to live, to exist as an *ex-sistere* (to be outside) on the one hand, and the man pressed in the dust of hard work on the other and handed him over to death (Gen 2.16–4.20). The original integrity and unity (identity) with God, creation, and nature gets broken.

The individual and social restoration of identity would therefore be a protection against the falling apart of the psychic personality, a safety net against the existential "thrownness" (*Geworfenheit*) of man in the sense of Martin Heidegger (1927/1962, §43, p. 255), a reinsurance against the loss in the world, a consolation against the (primarily repressed) knowledge of the transience of his actions, the superiority of death, frailty, illness, defeat, limitation. The idea of a secure, fixed identity offers protection against one's own psychological and physical fragility. A collective identity that seems unchangeable over time, established from traditions, rules, customs, and founding myths, makes it easier for individuals to hold together their identity—at the price of limited personal growth that is subordinate to the group. Where people experience themselves as weak, vulnerable, replaceable, and transient, they compensate for their fears by constructing a supposedly secure individual ("I am I") and collective identity: "I am my people," "I am my football team." Or: "I am my *Heimat*."

Habermas attaches central importance to this transition: "Collective identity regulates the membership of individuals in the society (and exclusion from there). In this respect, there is a complementary relation between ego and group identity (…). This suggests that we can infer from the ontogenetic stages of ego development the complementary social structures of the tribal group, the state, and, finally, global forms of intercourse." (Habermas 1979b, p. 111).

From existential need to threatening consequences- the conflicting aspects of identity
In his hope of overcoming reductive identity models, Habermas admits that he initially underestimated the connection between collective identity, cultural values, worldviews and norm systems. In a way, only a certain section of the culture and the system of action is important for the collective identity: "Individual members of the group perforce experience the destruction or violation of this normative core as a threat to their own identity. The different forms of collective identity

can be found only in such normative cores, in which individual members 'know themselves as one' with each other." (Ibid).

This shows the process of identity performance which, on the one hand, defines its "necessity", which is justified by many social scientists, but on the other hand also its dangerousness: the reduction of complexity, the simplifying compression of everything that defines personalities and groups to a core identity or an identity core, shaped by the requirements of tradition, myths, family, society. Fraying from this form of imprint brings greater freedom, but also demands greater effort from the individual in the self-assurance of his existence, in the self-assurance of his psychological and physical continuity, and in the self-reflection of his meaning. While Habermas postulates *universalistic ego structures* and *universalistic morals*, i.e. ultimately a universalistic identity (cf. Habermas 1979a, b, c)—after a profound criticism by Ingrid Jungwirth (2007)—most identity theorists face the dilemma of ultimately contributing to the mythization of identity through the question of identity: "Speaking of identity is necessary because it is a necessity. It is because it is so. The postulate of the inevitability of the 'question of identity' is based on itself." (Ibid, p. 31).

For Jungwirth, even socio-constructivist or de-constructivist concepts of identity ultimately only have the consequence—even if hidden—that "the concept of identity is retained, since one's speech is endowed with the power that narratives of origin have. It is impossible to speak of identity without metaphysical content" (ibid, pp. 32 f.). Using the example of the identity deconstruction by Stuart Hall (cf. 1990, 1996), Jungwirth shows how Hall opens the door for the concept of a necessary identity, which comes in again through the back door at the end. For Hall, in its basic concern, identity is only a fiction of naturalness and is constructed historically, culturally, and politically (Hall 1996, pp. 442 f.).The idea of an "essential identity" is an "imaginative rediscovery", which is opposed to breaks and discontinuities (Hall, 1990, p. 224). According to Jungwirth, Hall cannot completely free himself from the need to identify the individual with larger entities, since he speaks both of the "impossibility" and of the "necessity" of identity (Jungwirth 2007, p. 26). Ultimately, he adheres to identity as an irreducible concept due to its centrality in relation to agency and politics (Hall 1990, pp. 223–226). However, this also raises the question of how one can explain such a persistence of myths that emerge strengthened from every attempt at destruction if they did not actually have a comforting, calming, and reassuring—in this sense necessary—effect on people.

As "magic formulas" Lutz Niethammer (2000, p. 457) linguistically places the multitude of definitions for identity in the immediate vicinity of *Heimat*. For

Niethammer, the problem with the concept of a collective identity lies in its constitutive "delimitation from a non-identical" (ibid, pp. 625 f.). Thus the concept of a collective identity resides in "a tendency towards violence and – due to the religious charge – towards fundamentalism" (ibid). The uncritical establishment of a collective identity, as Jungwirth summarizes the recent social-scientific criticism, excludes that its misuse in the nineteenth and twentieth centuries had "murderous consequences" due to the supposedly natural lawfulness of inclusion and exclusion (Jungwirth 2007, p. 23). She criticizes Mead's basic identity theory for sociology because he developed it from the distinction between human and animal: "The rule of humans over animals and insects, from 'civilized' over 'primitive' societies thus becomes an evolutionary necessity." (Ibid, p. 129).

Identity building through crisis and Ego growth—the Eriksen-Concept
The concept of identity of the social and cultural sciences, as found in everyday language, was significantly shaped by the ethnographic studies of psychoanalyst Erik H. Erikson. With the essay "Childhood and Tradition in two American Indian Tribes" (Erikson 1945), he laid the first foundations for his national-character studies and his design for collective identity and personal identity (see Erikson 1950, 1968, 1994). Erikson distinguishes eight phases of identity development (cf. Erikson 1950, pp. 222–243):

- Infancy: The development of basic trust overcomes distrust and leads to the formation of unipolarity against premature self-differentiation
- Toddler age: Autonomy against shame and doubt leads to bipolarity against autism
- Play age: The initiative against feelings of guilt leads to game identification against (oedipal) fantasy identities
- School-age: the sense of work creates self-esteem and overcomes the feeling of inferiority—work identification against identity blockade
- Adolescence: The adolescent develops identity against the threat of identity diffusion (also "identity confusion")
- Early adulthood: intimacy against isolation creates solidarity against social isolation
- Adulthood: generativity against self-absorption
- Mature adulthood: integrity against life's disgust

For adolescence, Erikson outlines a more refined frame (Erikson 1968, pp. 128–134):

- Development of a time perspective against time diffusion
- The development of self-assurance against embarrassing identity awareness
- Experimentation with roles against negative identity choices
- Trust in your own performance against work paralysis
- Sexual identity against bisexual diffusion
- Leadership polarization against authority diffusion
- Ideological polarization against diffusion of ideals

As convenient as it is for insight and overview, the division of development into phases always involves the problem of simplification and categorizing, where it illuminates the possible development phases and constructs and fixes them to a certain extent. Some life tasks, such as those outlined by Erikson, are exclusive to adolescence and face people of different cultural contexts in other moments in life. A crucial point in Erikson's theory of identity is the gradual merging of personal and cultural identity into one unit, an "identity" in the true sense of the word, the highest level of which is "integrity," the maturation into an individual with personal responsibility and initiative. Erikson's famous sentence "Ego growth through crisis resolution"[2] (Erikson 1971, pp. 107 f.) sees the actual pacemakers of ego development in crisis-like life transitions. No matter how stressful the experience, how painful it can be to abandon familiarity if the crisis is managed successfully, the reward for an ego-growth, a more mature and autonomous personality, is waiting in the end.

An essential prerequisite for Erikson is the basic trust established in early childhood, which enables us to survive crises, give up the familiar, and turn to what is new. If the basic trust is damaged or has not succeeded, this leads to undesirable character and psychopathological developments. The individual does not grow in the crisis but gets broken or gets stuck with the risk of regression. With this, Erikson also establishes in his identity model a "healing origin," which is of course due to the history of life and is open: "To be a person identical with oneself presupposes a basic trust in one's origins and the courage to emerge from them." (Erikson 1964/1994, p. 97).

---

[2] This formula for development and identity formation is quoted literally in German receptions of Erikson's work, including Jürgen Habermas (1979b). However, in the corresponding English-language work "Childhood and Society" (Erikson 1950), the formula is not found in this form. Still, the interplay of crisis resolution in the respective phases of development and the formation of a social ego is addressed in several passages (e.g., pp. 229, 244, 371). The book's third chapter is entitled "The growth of Ego" (pp. 171–250).

Although Erikson cites the self-reliant autonomy of the personality as the goal of self-identity, the development of a mature self-identity remains dependent on social fit. The individual's sense of self-worth is not a subjective feeling for Erikson. Instead: The style of one's individuality must "match the equality and continuity of one's meaning for significant others in the immediate community" (Jungwirth 2007, p. 164). Habermas speaks of this as a "cognitive style marked as field-dependent," which the ego has to overcome going further in his emancipatory throw. The I has to replace the "field-dependent" style with a "field-independent style of perception and thought": "Autonomy that robs the ego of communicative access to its inner nature also signals unfreedom. Ego identity means a freedom that limits itself in the intention of reconciling – if not of identifying – worthiness with happiness." (Habermas 1979a, p. 94) With Erikson in adolescence, "ego identity" and collective identity have to be brought together to form a unit, which also underlines the importance of national affiliation for adolescents. In this way, identity becomes "a stabilizing factor" (Jungwirth 2007, p. 165) for social order.

And what happens if the basic trust develops not in a desirable way? In his studies of African-American children, Erikson's positive development model tilts to the opposite; the "black identity" (Erikson 1950, p. 217) leans toward "negative identity," as Jungwirth (2007, p. 190) criticizes. The circumstances in which black (and other indigenous) children grow up would prevent positive identity development (Erikson 1950, pp. 217 ff.). Just as all favorable development factors are united in mature ego identity, the identity formation process would also merge all negative parts into a negative identity. With such a literal black-and-white view, Jungwirth accuses "racial construction": "Erikson speaks concerning feelings of white children that the white is defined as the general, from which the black deviates (partly due to pollution), while the black people may think that the white was created by bleaching the black – in a comparison he equates white children with black people and also traps semantically: White is the result of cleaning/bleaching, the black of pollution." (Jungwirth 2007, pp. 196 f.) In Jungwirth's interpretation, being white is "civilized" for Erikson, whereas being black or Indian is "primitive" (ibid).

This critical reading by Jungwirth of Erikson's identity concept represents a ridge walk. Erikson's interpretations of white/black are psychoanalytically grounded and would be imaginative. The statement on "positive identity" refers to the worry that such is impossible under child-hostile circumstances. Because of his Jewish background, Erikson may also have been aware of racist patterns, so these were probably far from his mind. At the same time, it shows how quickly identity concepts can tip over into generalizing judgments, how close the step is from a social concern for the disadvantaged to their categorization as a lost group.

The critical issue of the identity discourse is the idea, widely accepted even beyond Jungwirth's critique of Erikson, that the individual's identity has to coincide with that of the group. The social pressure for the fitting between the individual and the collective to an unbreakable unity bears the idea of a closed We. The conformity of the individual style with social norms and values is a prerequisite for "positive identity," this inevitably casts a shadow over those who are outside the majority culture. According to Jungwirth: "Continuity and coherence of identity are based on conformity with norms. In this way, white normativity, the norm of binary genders, norm of male superiority, and heteronormativity in the speech of identity are discursively produced and justified as scientific norms." (Jungwirth 2007, p. 207).

Identity thus stands for unity with a *culture of the majority*, as shown in the subsequent studies on the ethnicization of identity using the example of South Tyrol. Behind culture as agreement shaped by majority instead of multicultural and transcultural plurality lurks a human image of non-difference. The apparent homogeneity almost inevitably excludes all those who deviate from the majority. From this perspective, identity becomes the exclusion of ethnic, cultural, economic, religious, social, gender, sexual, and any other diversity.

Beyond Gender, Skin Colour, Disability, Origin—Perception of Differences as Educational Challenge

The dynamics of exclusion and devaluation through identity constructions trace the postcolonial discourse. With the provocation "that the black man is not a man," Frantz Fanon (1967, p. 1) shows that everything associated with humans is a construct based on white ideas. The black person is the deviation from the white norm (ibid), which exclusively claims humanity from a racist perspective: "Rather, in a racist society, only white people are human, and people of color are instead Other to human, or beasts and animals." (Ibid) For Homi Bhabha, Fanon shows in this analysis that stereotypes are "not a simplification because it is a false representation of a given reality. It is a simplification because it is an arrested, fixated form of representation" (Bhabha 1983, p. 27), from which the people behind the stereotype are no longer released: "Stereotypes of savagery, cannibalism, lust, and anarchy [...] are the signal points of identification and alienation, scenes of fear and desire, in colonial texts. It is precisely this function of the stereotype as phobia and fetish that, according to Fanon, threatens the closure of the racial/epidermal schema for the colonial subject and opens the royal road to colonial fantasy." (ibid, p. 25).

An almost fatal dynamic is reversing from a positive stereotype into a negative connotation. From considering Jews as intelligent, diligent, and thriving, there was a small step towards racist profiling as malicious, greedy, and overpowering. For Fanon, the potentially positive connected metaphor of "black soul" would be nothing else than "a white man's artifact" (Fanon 1967: 14). In the same way, the stereotype of the wild, strong, and maybe sexually potent black man quickly turns into negative as a potential rapist, following the canonical equation 'black is equal dark is equal night is equal irrational is equal devil' (cf. ibid, p. 188; cf. Jungwirth 2007, p. 211). In contrast to the archetypes of C. G. Jung, the colonial collective unconscious for Fanon is the "sum of prejudices, myths, collective attitudes of a given group" (Fanon 1967, p. 188). According to Sara Ahmed (2006, p. 139), "for the black man, consciousness of the body is third-person consciousness, and the feeling is one of negation."

Hans Henning Hahn and Eva Hahn (2002, p. 48) write about the importance of stereotyping for the emergence of nationalism: "If we take the emotional loading of the stereotype seriously, then it is probably an admissible hypothesis that (real or imaginary) conflict situations are highly 'stereotypogenic,' i.e., represent the context in which generalized judgments and images quickly become emotionally charged stereotypes. It is less the real experiences of conflicts that have been fought than the emotions that precede them, accompany them, or emerge as effects (such as a martial defeat or victory) that shape the stereotypes. It is the fear of threats and not the real threat itself that the analyzing historian has to focus on."

With identity designs, differences are smoothed out and ironed out; discontinuities are bridged in favor of coherence and constancy. The subject creates unity through identity construction, which reduces its complexity and multi-dimensionality. This is not only a theoretical but substantial "crossing point" (Purschert 2012, p. 351) where postcolonial theory and queer theory meet in a body- and space-oriented phenomenology. "We 'become' racialized in how we occupy space, just as space is, as it were, already occupied as an effect of racialization." (Ahmed 2006, p. 24) Bodies materialize themselves "in a complex set of temporal and spatial relations to other bodies, including bodies that are recognized as familiar, familial and friendly, and those that are considered strange" (Ahmed 2000, p. 40). The white body occupies the world as it would be his world only, but more precisely, it is the heterosexual male white body. The patriarchal normative-heterosexual hierarchy tries to keep the male control over reproduction and discriminate for this the female and queer oriented gender.

According to queer research, at the center of this process, around which national, cultural, and political identity groups are formed, is the construction of a precise and uniform gender as the structuring principle of human beings. For Judith Butler, "Sex" is, thus, not simply what one has, or a static description of what one is: it will be one of the norms by which the "one" becomes viable at all, that which qualifies a body for life within the domain of cultural intelligibility (Butler 1993, p. 2).

For Butler, the biological gender is more than "a physical given" in the formation of identity; it is only created by reducing many possible genders to the opposite pair of men-women through the construct of the social gender. Margaret Mead already discovered in her ethnological studies in the South Pacific from the 1920s that there are cultures with multiple genders and those that recognize intermediate genders (Mead M. 1967, 2001). The Navajos and other North American Indian cultures recognize five or more social and three biological genders (cf. Lang 2006, p. 195). In Oman, the Xanith or Chanith are considered the third sex, anatomically endowed with the primary male sex characteristics, but socially classified as women due to their activities and preferred sexual practices. The decisive factor for gender assignment is not the genital equipment but the question of whether they penetrate or not (cf. ibid, pp. 201 f.). Bugis on the Indonesian island of Sulawesi distinguishes five genders. In addition to men and women in the traditional sense, there are anatomical women with male preferences (Calalai), anatomical men with female tendencies (Calabai), and the fifth gender of the Bissu, which can be anatomically male, female, or intersexual, as a combination of all genders and their clothing. What is important is that none of the genders are considered a deviation; the Bissu enjoys special reverence due to their mediating role between people and spirits (cf. Graham 2002; cf. also Röttger-Rössler 2000; Tol et al. 2000).

The ethnographic knowledge about the cultural condition of the supposedly safe biological gender also receives scientific approval. For Heinz-Jürgen Voß, who compares and brings together the philosophical, psychological, sociological, and scientific gender discourses, several genders can also be demonstrated according to biological theories (Voß 2010). Claudia Lang divides the supposedly compact "biological gender" into five subcategories: the chromosomal gender (XY for male, XX for female), the gonadal or gonadal gender (testicles or ovaries), the hormonal gender (gender-typical mix of "male" and "female" hormones), the internal genital sex (prostate or vagina, uterus, fallopian tubes), the external genital sex (penis and scrotum or clitoris as well as small and large labia) and the phenotypic sex (cf. Lang 2006, p. 68). If, for example, the hormonal gender already shows different "blends" and smooth transitions (ibid, p. 76), an ever more differentiated picture of sexuality emerges through ever more precise detection and determination of additional and new gender characteristics (ibid, p. 69). The supposedly objective biomedical models turn out to be "social constructs" (ibid).

Male-female is a rough distinction based on the most eye-catching gender characteristics, omitting all intermediate and mixed forms by hiding all other, more refined aspects. This leads to reducing the concrete person to a few characteristics and depriving them of their plurality and diversity. A similar narrowing of perception can be observed on all those lines of difference that create dichotomous constructs. Thus each exposes one half of the split reality to discrimination and devaluation: reason-drive, male-female, normal-disabled, gifted-learning-weak, Christian-Moslem, native-foreigner, black and white serve the bias.

One example that impressed me deeply and shows how even reflected and sensitive people are guided by dichotomous structures of perception is an episode on the fringes of a conference with a compassionate colleague. The woman had a more extended conversation with a colleague whose name she did not know. She then approached me and asked me the man's name in the wheelchair because it would be such an interesting man. Immediately she grabbed her mouth and added: "You see, what we criticize scientifically happens to us so easily. I could have asked you the man's name with the long hair and the handsome face. I fell for the very first feature that identifies him as disabled." Unfortunately, such a reflection is rare. It is more common and persistent to cling to the isolated discriminatory feature put over the whole person as if that person is only black or only disabled or only homosexual or only non-mother tongue. This abbreviation of the perception of the Other through the fixation of salient characteristics causes culturalist, racist, and ableist, sexist and sexual discrimination. How often do we address adults and educated people who have difficulty expressing themselves in our language, like children or idiots? This is precisely the mechanism of devaluation by defining the other based on that one characteristic that supposedly separates him from us—we are native, they are a foreigner.

Such dynamics concern dichotomous distinctions, such as skin color, language, gender, religion (e.g., through headscarves), disability, and how one dresses and performs, as Bourdieu (1977) explained in the Habitus Concept. With the idea of postmigration, critical migration research offers a concept for this purpose, to free migration itself from the dichotomy between the supposed normal state (with a population assumed to be homogenous) and a state of emergency (through migration movements). "Only when established patterns of thought are overcome can the entire field in which the migration discourse is embedded be rethought. In this sense, it is an epistemological turn, a break with the separation between migrant and non-migrant, migration and permanence." (Hill and Yildiz 2018, p. 7; cf. Yildiz and Hill 2014) This was pioneered as a research strategy by the group Transit Migration (2007), in the sense of a 'de-migrantisation' of migration research and a 'migrantisation' of society research (cf. Römhild 2014, p. 38).

Dichotomies are robust and tenacious, not least because they sustain existing rule orders. The Latin saying *Divide et impera*, divide and rule, was not accidentally taken up by the keen thinker of power Machiavelli (cf. Xypolia 2016) and applies to historical field commanders as well as to the younger European colonialism (ibid). The German poet Goethe found scientific recognition through his efforts to refine the perception of reality; consequently, he turned the principle around: "Divide and conquer! Good word; unite and lead! Better hoard." (Goethe 1814/1962).[3] Pedagogically spoken: The separation and subdivision of children, young and adult people according to sharply defined categories can only be weakened through a differentiated exercise in perception, also in the sense of habitus-reflexive education (cf. Vogel 2019). If we neglect this vital exercise in educational contexts, as it is the overwhelming case in politics, the concerned people—victims *and* perpetrators—remain in their prisons of categorization.

This is a crucial point for pedagogical reflection and action for diversity learning: constructing pairs of opposites, which serve to fade out and devalue diversity and create a pseudo-homogeneity, obeys the law of dichotomy. The discriminated characteristics are naturalized and subsequently established as absolute, without nuances, without smooth flowing transitions and blends: male or female, black or white, native or foreign, or belonging or stranger, *tertium non datur*. Dichotomies inevitably lead to the denial of all intermediate forms, but also—psychoanalytically speaking—one's commonalities with the Other. The dichotomous thinking, from which also Descartes' dualism springs and which is highly successful due to its complexity reduction, lies at the origin of strategies of domination through the creation of orders. In a pair of opposites, ruled by the law of either-or, one

---

[3] Goethe's quotation is translated in different ways, e.g., "Divide and rule, a sound motto; unite and lead, a better one" or "Divide and rule, the politician cries; Unite and lead, is the watchword of the wise." The translation by the author in this book is based on the following original German text: "Entzwei und gebiete! Tüchtig Wort; Verein und leite! Bessrer Hort." The play on words between word (as a reference to language as an instrument of dichotomy) and woard (as a treasure and in German also as a shelter) made the proposed translation seem more meaningful for our context. It is interesting to note that Goethe does not translate the Latin "divide" with "teilen" (divide in English) but with "entzwei" (cut in tow), which comes closer to dichotomy.

of the divided parts is inevitably in the position to exercise power over a subject. So the dichotomous gender roles establish almost coercively "patriarchal rule" (Larcher 2000, p. 46). For Judith Butler, the matrix of clear sexual identification constitutes "the subject [...] through the force of exclusion and abjection, one which produces a constitutive outside to the subject, an abjected outside, which is, after all 'inside' the subject as its own founding repudiation" (Butler 1993, p. 3).

In psychoanalytic theories, the dichotomy corresponds to the dyad, which needs the opening by the involvement of a third party. Triangulation would be the recommended way to cope with ambivalences wiped away in dichotomous thinking. An identity concept subject to the law of dichotomy inevitably leads to stereotypes that discriminate between belonging and not belonging, own and strange, higher and lower, noble and base, pure and dirty.

The division of human beings into exactly two and not more parts, which can temporarily unite, but are ultimately irreconcilable, is the subject of Plato's "Symposium" (Plato 2008). According to the myth, the gods punished the three sexes united in the spherical beings because "they tried to make an ascent to heaven" (ibid, p. 23). The holistic, three-sex man would have been too powerful for the gods. So they beat him in two, with variables in the gender distribution—male, female, male-female, so that the resulting halves desperately seek and perish:

> *"Zeus took pity on them and thought up another plan: he moved their genital organs round to the front. Up until then, they had their genitals on (what was original) the outside of their bodies, and conception and birth took place not in the body after physical union but, as with cicadas, in the ground. By moving their genitals round to the front, Zeus now caused them to reproduce by intercourse with one another through these organs, the male penetrating the female. He did this so that when couples encountered one another and embraced if a man encountered a woman, he might impregnate her, and the race might continue. If a man encountered another man, at any rate, they might achieve satisfaction from the union and after this respite turn to their tasks and get on with the business of life." (Ibid, p. 24)*

In mythology, the punishment (death, mortality) is alleviated by the grace of human creation. According to the high prestige of homosexuality in ancient Greece, physical love granted by the gods includes not only the creating love between man and woman but also fatherless love among men. One served for reproduction, the other as a higher form of love for Eros, work, and knowledge (cf. Plato 2008, pp. 24 ff.). But the division of human beings has been carried out cruelly; unity and identity can only be temporarily achieved through sublimation and makeshift measures. This shows the pattern of dichotomous division in either-or-schemes. Identity becomes compensation for the split subject on the symbolic level.

Just as good and evil are difficult for people to cope with as belonging together and are therefore split into good or evil, any sexual orientation beyond the binary social order creates worries. In this way, male and female parts are strictly separated by splitting off: the man has to be ultimately man, the woman completely woman. On the one hand, even Freud, who tried to justify bisexuality, unconsciously undermined his concept in which he praised the "most complete mental masculinity" of some homosexuals (Freud 1905/1962, p. 142).

The pattern that identity constructs recover while deconstructing them runs through the social and cultural sciences—as Jungwirth demonstrates in almost all identity models. Understanding strong motifs such as home and identity cannot be about deconstructing and rendering them harmless once and for all, but understanding deconstruction as an open process that never comes to an end since each deconstruction produces new constructions. The pedagogical challenge is not about conceptually abolishing identity or *Heimat*, which would remain an exercise in science. Pedagogy has to work with the underlying (and repressed or compensated) needs, making people prefer to establish themselves in a simple, reduced identity rather than a richer one to venture into more complex, but also more ambivalent, self-designs. The feminine in the male, the masculine in the female, assuming the foreign in one's own, would help to no longer consider social orders absolute—according to the myth of origin. In the optics of manhood that feels threatened, the price would be to give up the man's flight from his femininity. The gain would be to live a more multifaceted identity with parts also connoted as 'female'. Conversely, the recognition of 'male' connoted proportions of women would assign her more easily and naturally the place in public life that the women's movement had to struggle for worldwide.

In the optics of an identity that feels threatened, it would not be meaningful to project the hostile onto enemy images but to perceive it as a task in one's reality of life. As split-off parts, the feminine, the hostile, and the foreign can be projected into multiple destructive fantasies. A possible projection surface, which is visible in Tyrol, is the maternal, feminine, politically, linguistically, and culturally clear and undamaged *'Heimat.'* Referring to the initially quoted riflemen values, there should be no "mixing," no "multicultural" loss of meaning and value, no delimitation of identity as an "unmistakable way of life and attitude to life, based on tradition" (like the rifleman captain said, see above).

The following case studies confront these on the symbolic level hardened values with the actual experiences and conditions of young riflemen and marketers in their lifeworld.

"They Don't Know Where They belong" Case Studies of Youthful Identity Formation

Theoretical Introduction: About Problems with and Persistence of Categories

For presenting youthful identity designs in this chapter, a preliminary theoretical remark is necessary, which may qualify the accounts of the youngsters and the accompanying reading. Reflecting on youth in the context of identity formation, considering the aspect of ethnicization means stepping onto a multi-brittle terrain that is interwoven with structures solidified by discourse. The terms and their subjects and the objects behind them are fleeting: What is youth, when does it begin, when does it end, how does it define itself, in which lifestyles, and through what necessities does it perform? The delimitation of *youth* (cf. Böhnisch 2008, p. 30), both in a temporal-biographical and socioeconomic dimension, has become the standard of social and cultural studies. Young people's habits are now considered pluralistic and decoupled from a precisely discernable age phase: "The young people do not form a monochrome universe, but show themselves in deep breaks and segmentations and follow differentiated designs." (Merico 2007, p. 28). Therefore, research on young people also requires differentiated interpretative categories since the conventional "youth" container, which stubbornly survives despite being questioned, is no longer able to standardize the behavior and practices of young people's lifestyles.

Youth research also requires reflection on the position from which the speaking about youth occurs. Analogous to "The discovery of childhood" (Ariès 1962, pp. 33–49) and thus also "Conquest of the child by science" (Gstettner 1981), the discovery and associated definition of youth is often a story of attribution by definition (also in the sense of the Latin word origin *definire*, delimit/limit) and object formation. Ultimately, speaking about youth obscures the intent and domination of those who speak—usually the adults.

A similar dynamic of enforcement, stabilization, and—against all questioning—persistent assertion of an anthropological figure such as *youth* was already discussed in the previous section about identity. The insistence on researching, describing, and defining identity—as explained in the previous articles—creates in the first place and solidifies (metaphysically) the object to which it adheres even with critical distancing and liquefaction (cf. Jungwirth 2007, p. 26). The same applies to culture and ethnicity, which on the one hand, create opportunities for assignment, through which subjects can constitute themselves, and on the other hand, can be constricting and defining.

Ultimately, getting caught up in such attempts to overcome concepts is inevitable. The mere effort to dispense with terms like youth, identity, and ethnicity as order categories asks for thinking and writing about the impossible. Definitions can be delimited, but only at the price of new definitions. They need to be delimited again if they are not intended to establish a metaphysic of the described subject (youth, identity, and ethnicity).

Youth, identity, and ethnicity represent orders and constructions that convey a feeling of controllability of the objects we are talking about and thus a feeling of control over the world we live in. And this is precisely where the power of both identity and ethnicity lies at their crossing points beyond age groups, their potential (for empowerment), and their limitation (incapacity and attribution). Even with pluralization—in the sense of identities, ethnicities, homelands—it is impossible to escape this dynamic completely. Definitions in the plural open up thinking about the concepts but at the same time strengthen the figures of thought behind them through modernization and refreshment. On the one hand, constructing an ego that is *one with itself* and its peers, which through ethnical classification goes into a *collective Us*, has an authoritative effect. Self-esteem and belonging mobilize and protect the idea of a personal *and* collective self, which is brought about by differentiation from another person necessary for this. The clearer, safer, more solid one's self must be, the sharper and more insurmountable this Self must be differentiated from the constitutively necessary Other—so as not to blur the contours of one's self and not to unsettle it, with all the problems and extreme historical examples already described regarding annihilation of others (cf. Peterlini H.K. 2016a, p. 136). Even adapting the other to one's self is problematic since it also changes and creatively rearranges it in the adaptation process (cf. Bhabha 1996, p. 58). If so, the other must be assimilated entirely, unmistakably made equal to what is one's self, which ultimately also amounts to the annihilation of the other *as the Other* by erasing the differences and their quirks, as Byung-Chul Han (2017) thematized the "expulsion of the Other."

Ambivalences Between I and We: Unity with Oneself and in the Collective We as a Result of Dichotomous Divisions

The described processes of delimitation are associated with a process of assignment assumed necessary for almost all theories of identity. For example, as described in more detail in the previous chapter, Erikson viewed identity in his "national character studies" from the perspective of American nation-building (cf. Erikson 1974, pp. 60 ff.). As a German immigrant in America, he experienced the tension between migrant origins and uncertain new ways of finding one's home (cf. Wuttig 2016, p. 56). Erikson's ambivalence between principally possible individual openings and de facto collective closings expresses a problem of identity formation that is difficult to avoid. Identity functions, on the one hand, as a process of becoming and stabilizing an "ego" (Zirfas 2010, p. 11), i.e., the idea of yourself as a closed and constant *person/personality* (ibid). On the other hand, identity building requires the adaptation *to and* adjustment *into* offers of social belonging on the other hand: "A sense of identity means a feeling of being at one with oneself as one grows and develops; and it means, at the same time, a sense of affinity with a community's sense of being at one with its future as well as its history – or mythology" (Erikson 1974, p. 27).

This combination of personal and collective identity is based on the postulated harmony with the *history, mythology, and future* of a community and the effectiveness of national and/or ethnic offerings. In the present context, the *future* can only be meant as a projective idea determined by the other two factors with which the ego must be at peace, namely with the history and *mythology* of a community. Although potentially—and the diversity of youth (s) shows it—differentiated designs of *identity(s)* are possible, identity formation oscillates between creating an autonomous subject and its no longer mere autonomous adaptation to given social affiliations. In societies that define themselves—due to history, mythology, and imaginations of the future—as *national* or *ethnic*, other offers of membership are sidelined. Displaced in niche cultures, heterotopias, alternative scenes (cf. Foucault 1986), they are now only understood as abnormal deviations from a *true* identity in which the individual is *in harmony with himself* and *with the history and mythology* of the country. However fluidly theorized identity might be, in nation-states and nationalized ethnic majority-minority areas, it solidifies a *culture of the majority* based on the demarcation from the outside (other nations, other ethnic groups) and on suppression of internal differentiations.

National and Ethnic Identity Creation: The Example of South Tyrol/Italy

The European nation-state is a historically recent phenomenon born out of the national movements of the mid-nineteenth century. These were liberally inspired, supported by open-minded bourgeois circles and the young student community (Peterlini H.K. 2016b, p. 145). They coveted more democracy and rights for the people and were increasingly critical of the absolutist, mostly dynastically ruled state. In place of the dynasty, the common language of the people became the unifying identity motif—a late consequence of the printing press (cf. McLuhan 1962), which brought literature among the people and thus also upgraded the respective languages. Ultimately, the Habsburg dynasty, whose empire stretched from Italy to the eastern European north and south, collapsed precisely because of this development. But how could it happen that an initially nationally inspired liberation movement turned into nationalistic hardening? Germany, Austria, Spain, Italy reverted to absolutist, fascist orders; France struggled to fend off the fascist coup. The liberated European Eastern states found themselves in part likewise in a Marxist system of coercion. Partly already in the First World War, then even more brutally in the Second World War, the new European nation-states faced each other in cruel battles.

This is the dark side of the nation-state, forced by a canon of collective identity formation. Of the many possible identity characteristics, one usually comes decisively to the fore and becomes—in this case—the national cement. All other possibilities for individuals and groups to define themselves thus become secondary and partly even displaced. If the language becomes the determining identity feature for group togetherness, then the differentiation from other languages becomes very powerful. On the one hand, the nation-state seeks linguistic homogeneity internally, which was not a given in the first place, even in supposedly linguistically homogeneous states like Germany and Italy. One language prevailed everywhere, regional idioms were devalued to dialects, and the language of ethnic minorities was eradicated as far as possible. And to the outside world, the national "We" strengthened themselves through demarcation and ultimately also hostility toward neighboring national "We." Thus, for much of the nineteenth and twentieth centuries, leading European linguists considered it detrimental to character and health for a people to speak several languages. Multilingualism disturbed national identity.

Once again, the psychoanalytically described meaning of identity and identification can give hints for understanding. According to Erdheim (1984), the need of individuals for protection and consolation in their existential distress promotes a double strategy, expressed on the one hand in identification with a (mostly transfigured or overestimated) light-figure, dominant figure, or overarching idea, and on the other hand in gathering together a collective We (cf. ibid, p. 25). This double entendre makes it extremely difficult to question the identification figure and the group cohesion. The individuals avoid anything in their need for protection that could bring them into conflict with the imagined unity (cf. ibid, p. 38). Such a dynamic explains why social inequalities are absorbed by nationalist movements and shifted to national levels of conflict, where the aim is not to combat social disparities but to blame minority groups that act as an enemy. Emphasizing and demanding social interests would at least open up cohesion within national communities in favor of negotiating socioeconomic claims. The national group unity would be threatened and restructured along with socioeconomic categories. Thus, it would be risky for individuals to make social differences salient. They could thereby contrast with the nationally defined large group and risk the punishment of social exclusion. From an anthropological perspective, social exclusion is equivalent to social death. Shifting social problems to national debates instead exonerates the individual. Depending on its status, the individual continues to be socially disadvantaged but can console itself by being lifted in a national We and project all the social frustration on the national Other. This makes the national unity an empowering factor for people, with the prize of obscuring social interests and inequality.

Precisely because people are not identical with themselves, but as decentralized subjects are constitutively dependent on recognition by others (cf. Hegel 1807/1977, pp. 114 f.), they long for unity (in/with themselves) and belonging (to supposedly the same). The national We promises this particularly powerfully because it covers the dividing social differences within society and helps individuals imagine self-confidence and security. Promises of oneness can finally be fulfilled maximum to a small extent. Still, at the cost of suppressing individual and collective differences, it does not disenchant the national offer's effectiveness. Still, it strengthens it through fear, anger, the uncertainty aimed at constantly new enemies.

The Project Study: Identity Pictures of Young Tyrolean Riflemen and Marketers 1997–2009

A clear example of nationally and ethnically founded identity formation is the Association of South Tyrolean Riflemen. The already presented organization sees itself as a movement to protect the German and Ladin language minority in the ancient Habsburgian crownland and the more recent Italian province of Bozen-Bolzano/South Tyrol. The corps, which has long been ridiculed as an old and outdated homeland protection troop, has become increasingly attractive to some South Tyrolean youth through constant rejuvenation in recent decades (Peterlini H.K. 2011, p. 158).

In a project about young members of the riflemen in 1997, randomly selected young riflemen and marketers between the ages of ten and 25 were interviewed, accompanied on public occasions/marches, and visited in their private lifeworld. In a second step, the same young people were contacted again twelve years later, in 2009, and asked again about changes in life events, occupational and educational paths, changes in location, meaningful encounters (Peterlini H.K. 1998, 2011).

The project aimed to better understand youthful identity-building between a politically ethnocentrically defined critical culture and plural lifeworlds. Through dense descriptions (Geertz 1973), case histories were worked out. Behind solidified ideas of the (inflated) Self and the (dangerous) Other appear nuances and blurring.

In one of his first statements, a young rifleman said that "people who do not know where they belong" and "people who refuse to be Tyrolean" definitely do not belong to his *Heimat* (Peterlini H.K. 2011, p. 35). Belonging would only be possible for those who "submit to the Tyrolean virtues, customs, and conventions" (ibid). He considered "the interference and intervention of strangers in our country" the most significant problem for South Tyrol in 1997, 25 years after South Tyrol had become a model for protecting minorities through an autonomy statute (ibid). At the same time, he answered many questions in a differentiated and thoughtful manner. Another young rifleman mentioned a possible future problem for South Tyrol "that there could be another war with Italy." When asked astonished what he meant, given the fact that there was peace, he said: "But someone could come [...] that South Tyrol must defend itself against Italy." (Ibid, p. 63) The portraits below, for which the young people also made themselves available as young adults, show the processes of identity formation between deep-seated traditional patterns and individual openings and their prioritization. The young people's ideas from the first phase of the interview are analyzed first to reflect on them ten years later and reflect on changes or preservations.

In addition to psychoanalytical models, the methodological approach aims at a study of the lifeworld. Fundamental to this was the concept of system and lifeworld (*Lebenswelt*) by Jürgen Habermas. Just as in psychoanalysis, it is also essential to handle such models with caution for not imposing a theoretical concept on those concerned and overlooking the specific nature of their situation. The Habermas concept is a model of analysis: What are we talking about now when lifeworld or are the systems powerfully casting their shadow on it?

A few brief notes on this: Lifeworld in the sense of Habermas does not mean everyday practice, which would be a facilitating gift for this work. Therefore, the term needs to be adapted to use "Lebenswelt" and "System" as aids to understanding concepts of identity and belonging that are required for life and misused to reject life. The model seems to be tailored to South Tyrol: at the level of the lifeworld, people of different histories and origins live well together, they negotiate their problems, somehow get along with each other, even create a colorful, creative togetherness, for example, when colors, smells and languages mix lively in the fruit-growing area in Bolzano and nobody would think that this is a land of conflict. On the level of systems, reflected in political and mediatic discourses, there is often fierce competition between cultural and linguistic groups for symbolic dominance, for instance, in the dispute over monuments, place names, positions of power.

On a theoretical level, the lifeworld is for Habermas the unquestioned ground of all circumstances and the unquestionable framework in which people face the problems they have to overcome. The crucial point is that these lifeworlds, and also their everyday practice, only become accessible through problematization: "The horizontal knowledge that communicative everyday practice tacitly carries with it is paradigmatic for the certainty with which the lifeworld background is present, yet it does not satisfy the criterion of knowledge that stands in an internal relation to validity claims and can therefore be criticized. [...] It is only under the pressure of approaching problems that relevant components of such background knowledge are torn out of their unquestioned familiarity and brought to consciousness as something in need of being ascertained. It takes an earthquake to make us aware that we had regarded the ground on which we stand every day as unshakable." (Habermas 1987, p. 400) Such challenges also reveal the most important quality of the living environment: At this level, people are capable of "communicative action" (ibid, p. 403), which does not necessarily be conflict-free, but it enables solution-oriented negotiation processes.

In communicative action, people can agree on their demands and ideas about how they want to shape community life—even if it is through conflict resolution. While the living world is thus accessible to "communicative action" through the "becoming problematic" of its sub-areas, in the "systems," one dominant steering medium prevails in each case: for example, money in the economic system or power in the political system (cf. ibid, p. 165). At a workplace, "communicative action" can be possible between boss and employee, such as negotiating ideas about work processes, work necessities, and even wage issues. But as soon as the level of the "system" is reached, in which "money" is the steering medium, communicative action is replaced by "strategic action" (ibid, pp. 87 ff.), which is success-oriented and geared towards winning or losing. Here, there is no claim to truthfulness in communicative action. The participants seek a communicatively negotiated solution; at best, truthfulness is faked to manipulate the other party and achieve one's own strategic goal.

Grey areas are possible. For example, Habermas sees the mass media, on the one hand, usurped by the steering medium money (through advertisement and sale pressure). On the other hand, they would also amount to communicative action. Thus they oscillate between authoritarian and emancipatory potential, between the influence of the advertising industry and the culture of resistance in editorial offices. The same actor can oscillate between the poles: The employer, who tries to overcome a crisis with his employees, can act communicatively; as part of a system in which money decides on survival or bankruptcy, he will be strongly conditioned by strategic action. "Systems" tend to colonize the "lifeworld"; to wrest the lifeworld from the system requires the emancipatory commitment of citizens.

Another example: a playground that parents design together or that the district council has set up is a part of the lifeworld. There will be discussions about which equipment is more suitable, whether grass should be sown or lawn mats laid or whether tree bark is a better soil. The same children's playground can be colonized by the system when it comes to whether the "local" children can still play happily in this playground because so many migrants also bring their children there. This, too, would still be an area of the lifeworld if a dialogue were to be established among parents and with the neighborhood council through the problematization of problems if positions—also conflictual—were to be exchanged if the factual issues were to be negotiated communicatively. But it is more likely that it will very quickly become a matter of who loses face, who wins, who has to give way, how the district councils will look to their electorate, how a citizens' committee can be appeased. Ideology will take the place of conversation; the system with its strategic laws will push the questions of lifeworld into the background.

The steering medium "power" for the system of politics ultimately means: one is the subjugator, one is the subject. If I don't want the others to subjugate me, I can't let them arise—losing or winning, being on top of being oppressed as either-or. While people in the lifeworld can achieve understanding and compromise even on political problems through communicative action, on the level of the system, knowledge through listening and telling is impaired by the medium of power. This can help us understand why German and Italian people in South Tyrol work and do business together almost without problems but find themselves in a frontal position against each other because of a dispute about historical monuments. Playgrounds and kindergartens are designed beautifully and used by all groups but become the driving force of political debate about whether they should be ethnically separate. People who get along well with foreigners in their lifeworld can nevertheless hatefully complain about migrants when the discussion concerns the political system.

In this sense, the Habermas model of system and lifeworld was method-
ically used in the project. Where do *Heimat* and identity show themselves
well-grounded in the lifeworld? Where are systems of order superimposed on
it? Can the ambivalences between *Heimat* as lifeworld and political, economic
appropriation be recognized? The following selected case studies of young people
in the study attempt not to give unambiguous answers but to leave interpretations as open as possible. They each take two snapshots of the work with the
young people at twelve-year intervals. The first momentary pictures were taken
in 1997/1998, the second in 2009/2010.

Sigmar Decarli[4]

Molina is Down There

Sigmar Decarli, Altrei, Rifle Company Altrei, eleven years old; Interview on
August 30, 1997, in Goldrain; in-depth discussion at the parental farm in Altrei
on November 27, 1997.

> *"The Italians ... they are ... they were cheeky during the war ... [...]. Because the
> Austrians have made a pact that they don't attack, and they attacked genuinely. Because
> they wanted to come ... meant they could come, just come up and conquer us. [...]"*

Don't you think that the Italians, now I don't mean the state, but the soldiers who
fought, that they were poor devils because they had to fight and die for it?

> *"Nah ... then they wouldn't have had to attack us."*

But the soldiers themselves didn't decide that; the governments decided that. Do
you think it was also the Italian soldiers' fault?

---

[4] The case stories were not anonymized, which was a deliberate breach of a taboo in social
science work. The interviews in 1997/1998 were carried out for an exhibition; the interviews
and their publication were carried out with the permission of the parents in addition
to the consent of the adolescents. In 2009/2010, all interviewees were of legal age; subsequent
anonymization would not only be absurd given the publication in the exhibition twelve years
ago, but would turn the interviewees who were motivated to engage in the conversation and its
publication into anonymous research objects, even though they were and are equal conversation
partners with a first and last name. Refraining from anonymization also increases responsibility
in dealing with the statements because the right of objection of the aforementioned is
not overridden by anonymization. In contrast, statements made by a young marksman
at the time who was no longer satisfied with the recent questioning were presented anonymously
and inevitably in fragments; the statements of a young person who died after the first survey
were also anonymized.

"They just said they were attacking ... without anything ... and then they called the riflemen together."

That was in the war. Do you think the Italians are still dangerous today?

"Nah."

No?

"Nah."

Why not?

"Because now it would be far too dangerous, a war..."

And otherwise dangerous that they take away the autonomy, the country?

"Nah."

Then why do we have to defend our *Heimat*?

"We always have to do it, though. And especially when there is a war. [...] Yes, even if you ridicule the riflemen, you have to defend them. Yes, and say that they are old-fashioned. Yes, they are, but that is a memory, the shooters, and everything will always remain. The colleagues dress up more, totally long costumes, really long. I don't like these modern things."

In the narrow hallway of the old farmhouse, a picture of the Tyrolean hero Andreas Hofer is emblazoned on the wall, next to a rifle target. The corner bench in the kitchen is new, made of solid wood, "even if it looks old," as the mother apologetically runs over the wood, which children's shoes have worn down. It smells of firewood and lukewarm milk.

The Decarli family's farm is remote, even remote for the Altrei, one of the last houses in the hamlet of Guggal, on a slope down into the former Ladin, now Italian, Val di Fiemme, a continuation of the Fassa Valley. "The last village" was the title of a Bavarian movie that disturbed the local people in the 70s—"we were portrayed as if there were only old things in our village, old people, old customs, old houses, nothing new," says Sigmar's mother. When the 1997 Mafia film by Felix Mitterer was shot in Altrei, the village protested in advance of a new disparagement as a South Tyrolean forest town; Sigmar played along with his school class as extras.

Altrei is a borderland, looking into the Italian-speaking Val di Fiemme but hanging on the German South Tyrolean lowlands. During the First World War, the front line resided here, one of the most embattled in the Dolomite War. The older names have Ladin sounds (Antereu, Anterew, Antrew, Anterü, Antreui, Altaripa). They derive from the long-rooted Italian name Anterivo, which did not emerge from the fascist forced translation. Until 1779 it belonged to the gentlemen Enn and Kaldiff residing in Montan and Neumarkt, not to the "Magnifica comunità" of Fiemme, a particular form of administration for common land ownership in the Italian area around Cavalese. In 1779, the Princess of Tyrol and Empress Maria Theresia negotiated a land swap with the Bishop of Trento. She got the wine village Tramin, which had belonged to the Bishop of Trento until then—to the annoyance of the locals. For this, she ceded Altrei to the "magnifica comunità". So it was a back and forth up until its recent history.

The language limit drawn here would have to bear the expansion cracks of an often tough neighborhood. But it runs peacefully and invisibly through the woods. "Over there" has always been *Welschland*, Italian land. But "over there" is not separated from another "over there." Forest paths lead from Altrei to the Italian village Molina as from Altrei to the German town Truden. To get to Altrei from the German-speaking villages, people by car have to pass the Trentino for a few loops. A few years ago, Altrei could only be reached by telephone with a Trentino area code until they were finally connected to the South Tyrolean local dialing network after protests of many years. Also, this seems to be outdated trouble in the age of smartphones and disappeared phone books.

But nothing seems to have changed in the village's character: at the 1921 census, the first after the connection to Italy, Altrei had 420 inhabitants, including nine Italians. In 1991, at the last census to date, there were 400, including almost 50 Italians. The increase in Italians at the most extreme language limit is far from the national growth.

The Decarlis have the history of the lowlands written in their family tree. The farm they recently moved to was called "Trenta" after an Italian owner, now it has been given the old name again: "Wasserle-Hof" (farmstead of the little brook). Eight cows, one horse, several hens. The Decarli family puts a pot of milk, sometimes two, at the intersection near the farm every day. The horse, a 17-year-old *Haflinger*, is called "Kugel" (ball) because it is so round. "We are all fat," says the mother, and Sigmar explains: "The dad is fat, the mom is fat, and the dog is fat." There is a milk calf in the barn. It's called "Kevin," the mother cow is called "Jenny." The calf sucks Sigmar's finger as he strokes it. When asked which animal he connects with *Heimat*, Sigmar doesn't name Kugel, Rex, Jenny, or Kevin, but the deer, "who is out in the forest, whom Tata has been trying to shoot for a while but has never caught."

Sigmar's father Erich comes from Laag, the Neumarkt faction down in the valley. In that area, the language border runs so blurrily through the village that the Italian government wanted to prohibit children of mixed-language families from attending German school to prevent the supposed Germanization of Italians—the reverse assimilation. A Karl Decarli was the fire brigade commander of Laag as early as 1933. Sigmar's father, Erich Decarli, was the rifleman captain in Laag before taking over the farm in Altrei. Now he is an honorary commander in Laag and is considering joining the riflemen company of his new home Altrei. Sigmar is also with the riflemen, above all, "because of the Tata," says the boy. Sigmar's mother Friederike is a Lochmann. The name Lochmann appears in the first documents about the inhabitants of Antereu; in the rule of Enn, the Lochmanns are listed among the family lineages with real estate.

"FL," the initials of Sigmar's grandfather Franz Lochmann, are hewn into a boundary stone in the meadow behind the house. Franz Lochmann was one of the German mayors used by the Nazis after 1943 to replace the fascist Podestà, the local prefect installed by Mussolini. The journalist and historian Claus Gatterer described them as "no docile tools of the national socialist-oriented association of South Tyrolean." They would have been "respected citizens and notables who once again brought solid and economical administration into the parish rooms" (quoted from Fontana 1993, p. 377). The South Tyroleans accomplished a balancing act more or less flexibly when they enjoyed the "liberation" by the Nazis from the fascist rule in 1943 and felt the worse evils as more compatible with the nation.

Before the Decarlis took over the "Wasserle-farm," they lived in Laag. Sigmar hardly knows any difference. Here he has the horse and the forest, down there it was ice cream every day during the summer, often up to six times a day: "from Uncle Maurizio, from Grandpa, from Tata, from Mom, it was adding up." In Guggal there is ice cream only once a week when the ice cream truck comes.

At the first meeting, Sigmar had said that he was not a very modern boy, not as stylish as the others who wear "totally long clothes." He feels older and "a little bit braver" in traditional costume—and also more old-fashioned, but he stands by it. He wants to stay with the riflemen until he is old. What did he like so much about the riflemen? "Everything, just everything." Sigmar was wearing a peaked cap and T-shirt as he said this, but not that long. Two months later, he wore sneakers with loose strands and pierced earrings when we visited his farm. He copied this from the "big ones," not in Altrei, but in the city above: "I liked it … a little bit, to show off at school. Someone said that you are gay when you have earrings, but that's not true … only if you wear them on the right." When asked which movies he can think of as *Heimat* movies, he says: "Those with old people who wear *Lederhosen*." The leather pants are also the most important garment of the traditional rifleman costume.

He has another pet, a Tamagotchi, the tiny virtual being of the nineties that has to be constantly looked after and nourished. Otherwise, it will leave the cyber world with an alarming beep. Sigmar's Tamagotchi is a dinosaur who once died because he forgot to put him to sleep. During his death struggle, he "took two dumps" in agony.

Sigmar enjoys helping his father on the farm, usually in the evening, throwing hay down a hole from the barn into the stable. He prefers that over going to school. The school is three kilometers away from the farm. If he has to serve in the early mass beforehand, Sigmar gets out of bed at 6 a.m. at the latest. In the Italian village Cavalese, beyond the provincial border, they offer a swimming course for 150,000 lire (about 75 Euro) twice a week from autumn to Christmas. Sigmar has been going there for six years, even if it is starting to seem boring to him because he is by far the best. In the South Tyrolean village Auer, he attended a shooting course with the riflemen; 88 of 100 possible points was his best result. When he transfers from elementary to middle school, he has to take the school bus to Neumarkt—which means getting up earlier and coming home tired.

The former remoteness did not prevent Altrei from leading a famous son in the village chronicle. Johann Baptist Zwerger, who was later appointed prince-bishop of Seckau in Graz by the Salzburg archbishop and built the cathedral there, was born in 1824 in Guggal, out of all places, where Sigmar now lives. Maria Veronika Rubatscher (1928) immortalized the figure in her story "How the dwarf Hansele became a giant."

Sigmar, just shyly chewing questions on the corner bench in the kitchen, becomes a giant Hansele outside. He jumps on the bareback horse, scratches its mane, and gallops across the meadow. He scrambles around on a huge boulder, carrying his father's telescope for hunting. "Have you ever climbed such a high stone," he calls over. "These are the zetas," he points to a shrub, confessing that he has tried smoking the dried branches before. On the narrow forest path, a salamander is looking a little stiff for a place for hibernation. Sigmar knows the areas where the best mushroom species grow. He not only recognizes the jay by its feathers but also by its flight.

Sigmar has a brother, Karl, and two sisters, Erika and Birgit. "One child died," he says. He is the oldest. His favorite subjects are "Environmental and local studies ... and history". At one of the old theaters of war, the Mangen Pass, Sigmar Decarli camped with the riflemen for the first time. He mentions the place as the one with which he has the strongest connection to *Heimat*: "Where we camped with the riflemen for the first time." With the family, he went to an old Austrian tank factory Gschwent, a terrible scene of war on the mountain front against Italy in the First World War.

The forest path on which the fat dog follows him leads to Trentino. Down in the valley, you can see Lake Stramentizzo. The Avisio river flows peacefully.

Somewhere up there, you get to Molina. No boundary to see. Again the question: what comes to mind about *Heimat*? "Andreas Hofer … the Romans." What is the problem of *Heimat*: "A war," says Sigmar, "that would not be nice." But there is no war. "But someone could come …" Why? "That South Tyrol has to defend itself against the Italians."

Sigmar is happy to lift the bus that takes the children from Altrei to Cavalese for swimming. His father works on wood on the slope under the street and waves up to us. Sigmar would like to become a goldsmith one day. He thinks he will probably have to go into the city for this.

Trying to Get Along with Everyone

Summary description 2010; Interview on September 26, 2009, in Altrei at the parents' farm; Sigmar is 23 years old; Sigmar is still with the riflemen, in the range of a first lieutenant, but is considering to leave.

> *"[...] because the main commander does too many things that don't sit well with me ... and that is not me ... I don't like what they do. For a while, it's all well and good, but they make too much fuss about everything. One could talk freely but not always demonstrate and do things. [...] The fascist's victory monument in Bozen, they want to remove, is a little insulting, that's right, it's a little insulting, but ... it is part of South Tyrol and ..."*

> *"There are many who are against the Italians and all. I got to know a lot of Italian colleagues at 16, 17 [...] and as soon as you are with Italian colleagues, it is said, aha, you are with Italians. That made me distance myself."*

> *"Honestly ... I see Heimat this way: I am at home in Altrei, and that is my Heimat, I am fine there. In other places I would never stay long because ... [...] In Altrei you know all the people, have all the colleagues and everything, that's why it's your Heimat, and it's also nice, let's say, from the landscape."*

> *"Foreigners? They would have to be better integrated, to be honest, because ... they form groups, in Bolzano, for example, hostile groups... really hostile. They often ran after colleagues of mine with a knife and such, and always in groups, and that would have to be prevented. If they were more with us, that might change because if they are always with their kind, they still think the way it was with ... in their countries. Then they would understand that they cannot afford to commit more violent acts and such because ... that's not so common for us, I would say."*

> *"There are certainly problems with ... firstly, for example, because the Germans learn too little Italian and understand nothing if you go further up north, and this leads to many problems. Because if you go to Meran and up in this area, they don't speak Italian well. [...] The school alone does not achieve that... Honestly, in school, middle*

school, I always had a D or F in Italian. Then I got to know my colleagues during my
vocational school, and from then on, I was always at a B, almost higher."

At the re-encounter in October 2009, Sigmar is 23 years old. He is easy to
find, still lives in the same place with the same phone number. The mother
willingly provides information on where to find him during the day, at work,
a mechanics company in Neumarkt, the main town in the South Tyrolean lowlands in the valley. Sigmar can still remember the interview twelve years earlier
and spontaneously consent to a follow-up interview.

His entries on community sites like Facebook and Netlog gave a certain pre-
view of how he might have changed. His profile picture on Facebook shows
him in a photo that has been graphically changed and slightly modified with a
casually hanging cigarette in his mouth. In his portrait, he mentions liking the
following music: "DJ Rudy Mc," "South Tyrolean Musicians," AC/DC; Italian
portals can also be found under the "Fan Pages," for example, one for the protection of defenseless animals. The photo on Netlog is stridently modified: red hair,
a sharply cut face, sparkling tiger eyes, flared jacket collar, futuristic background.
He chose "kompfi" as a nickname. Excerpts from his self-portrait state:

- About me: Who can, should do;
- Interests: sports, cars, television, nightlife & going out, computers; films;
  sexuality; internet; music;
- Favorite car brand Lamborghini; own car brand: Ford. Favorite motorcycle:
  Ducati;
- Cigarette brand: Marlboro, trying to stop;
- Alcohol: so-so;
- Favorite cities: Prague, Rome [written in English];
- Favorite books or authors: Eragon;
- Radio station: South Tyrol 1.
- The best thing that has ever happened to you: won a trip to Rome!!!

At the agreed meeting time, Sigmar waits in front of his farm; a pot-bellied
pig ponders before him. It's called Rudi. Rex, the hybrid between shepherd and
Berner-Senner, is already quite old at 15, but he still runs five to six kilometers
behind the car every day; in the past, you had to drive 60 km/h to keep Rex busy,
now it's enough if you drive 45 km/h. This is the only way to keep the dog fit
because you cannot let it run free, then it will go after the deer, which the father,
a hunter, will not tolerate. A second dog, a mixed breed of uncertain origin, is

trained for hunting, "Waldi," a female. And the sisters have two birds, "a kind of parrot."

There are no longer any farm animals on the Wasserle-farm in the Altrei hamlet of Guggal. On the one hand, this corresponds to the decline in livestock farming in the South Tyrolean mountain region, too complex and, under a specific amount of stock, too unprofitable. Hardly a South Tyrolean livestock farmer can live on less than 20 cattle without a side job, and with sideline work, barn work costs many people too much time and effort. Just above the bottom of the valley just 12 years ago, the fruit line has almost moved up to the Inn "Pausa" below the San Lugano Pass.

The Wasserle-farm's location is too high for fruit and wine, and with eight cows, they were unable to make ends meet. So Sigmar's father simply gave up the dream he had associated with moving from his hometown of Laag to his wife's hometown—to be a farmer. For the mother, as she explains after the interview, the conditions imposed by the EU are to blame for the fact that agriculture has been abandoned: "We would have had to build a dung and surf pit for 40,000 euros, we would not have managed to gain that back in 50 years." Faithful to his passion for hunting, the father took a job as "man for everything" at a company for hunting and fishing goods in Auer, back down in the valley with its "mixed-language" villages.

When asked about his childhood dream to become a goldsmith, which Sigmar had almost forgotten, he only responded to it when he said goodbye in front of the open car. That didn't work out, he says. However, the mechanic profession aligns with his passion for engines and technology. He is not satisfied with his job only because he earns too little. He simply doesn't have enough to save with the many kilometers he drives to work every day with the meager wages. He travels 20,000 km a year to and from work, he has now negotiated 1300 euros a month with his boss, but with more than 300 euros fixed expenses plus refueling, there is not much left for him; he also has to give something to his mother. He has heard that the people in the box on the highway collecting the toll get 2000 euros. He would almost prefer a job like that. Thanks to night shifts, you make good money and have plenty of free time.

In the interview, Sigmar has a sluggish way of talking; he hesitates repeatedly, pauses for thought, then follows precisely his train of thought; mostly, he lacks the words, cannot express what he wants to say. He knows that pretty well. He struggles to utter two words, both of which have to do with his consistent criticism of the riflemen in the matter: "Main commandant" almost does not want to roll over his tongue, it is a strange, unusual word structure, generally speaking, one speaks of the riflemen from the "state commandant"; the exaggerated marches against the "monu... uh ... ments" give him a similarly striking problem when pronouncing. On the other hand, he answers the question of what symbolizes *Heimat* for him without hesitation: "The mountains! The mountains come to mind quickly".

As in 1998, the conversation takes place in the kitchen of the Wasserle-farm. It is almost unchanged, a few kitchen appliances have been replaced, and the school supplies have disappeared. When his sisters come in, Sigmar asks calmly but firmly: "Wait outside until we are done." When asked about the siblings, he lists them in order, not mentioning as he did in 1997 that one has died. Sigmar was born in 1980, followed not until eight years later by Erika (1988), then Birgit (1989), and Karl (1992). He attributes it to the fact that "there are many siblings" that everyone in his family would also respect other opinions, "you get used to it."

Sigmar loses his train of thought only once when his mother comes in and, briefly, joins the conversation. The question had been, what did he think about when it comes to the "oppression of *Heimat*"? Sigmar replies: "... there are already many people who are oppressed and so, especially the foreigners who are oppressed, but they do not belong to *Heimat*." The answer was phonetically difficult to understand. When asked whether he thinks foreigners are oppressed, he affirms: "Yes, they are oppressed a lot." This is in line with Sigmar's own experience with foreigners; he always got along well, only his friends were sometimes threatened with violence. However, when describing the attacks, he remains vague and relativizes them with the belief that foreigners should be better integrated.

When asking further why foreigners would "not belong to his *Heimat*" for him, the mother comes into the room because she wants to collect the bread for the cattle drive but then sits down at the table. Sigmar suddenly answers with irritation: "They are not being properly integrated, and that is ... Please ask the question again." He changes the subject to the new keyword "oppression of *Heimat*":

> *"Oppressed ...? Let's put it this way, you can't say your opinion. If you say something about Austria, you will quickly be portrayed as radical because we are allowed to, for example ... My cousin had the South Tyrolean eagle on a shirt, but the school banned it because it is against the Italians ....I didn't quite understand it because English people who have the English flag on tops aren't ... It's a bit of oppression ...".*

The conversation gets a twist: The mother tells how Sigmar's younger brother Karl was no longer allowed to school with a shirt that said "Loyalty to the Land Tyrol" (*Dem Land Tirol die Treue*). For the first time, Sigmar now raises his voice and becomes louder. He confirms that many guys could sing this song better than some Germans in the discotheques in the Italian neighborhood villages, reconciling the patriotic hymn with his attitude. The mother leaves the room, comes back soon, and starts to wash the dishes. When she takes part in the conversation again, Sigmar says that the riflemen are doing good despite criticism. Again, he defends the riflemen and the traditional *Heimat* concerns in the mother's presence.

When there is an agreement between son and mother, Sigmar speaks louder and more straightforwardly. The mother criticizes that some riflemen groups refused to salute the political grandstand at the Innsbruck State Festival parade for the 2010 commemoration because the politicians are doing too little for self-determination (which means secession from Italy). Sigmar agrees; yes, that was "disrespectful." Sigmar also sticks to his fundamental attitude with sensitive questions and is supported benevolently by the mother (the "more radical" son looks more like the father). Still, the accents sound different: the criticism of the traditional understanding of *Heimat* is milder, the occasional approval of the riflemen stronger, such as the need to protect the German language, where he previously believed that the real problem was the dwindling knowledge of Italian among South Tyroleans outside of multilingual centers and areas. He no longer addresses the issue of foreigners in the presence of the mother. Discomfort caused by "strangers" only comes up again—in a completely different form—through the mother's grim expression of displeasure at the many tourists ("strangers") who invade the forests so that it is no longer nice there. Sigmar appeases; there were hardly any mushrooms this summer.

Unlike in 1997/1998, in 2009, Sigmar is no longer a child but a young adult. His views have become more independent of the traditional understanding of *Heimat* in which he grew up—as the oldest child of a riflemen family. When reading the interview from 1997, it is noticeable that Sigmar usually answered vaguely when asked about his *Heimat* views. He was more assured if he could rely on what he had heard at the riflemen or during the excursions to the theaters of war. Then he knew relatively well that Austria had a pact with Italy before the First World War, which is why the Italians were "cheeky" when they were fighting South Tyrol. At the time, he did not consistently accept the argument that the Italian soldiers themselves did not like to go to war—they remained caught up in Sigmar's childish image of the enemy "Italians" but influenced by traditional adult knowledge. When asked about his previous positions twelve years later, he says of his own accord that he had formed his opinion at the time about what he had "picked up": "But when you get older, you start to think for yourself…".

According to Sigmar, the following moments were influential for his learning and thinking processes:

- When he was a middle school student (in his adolescence), he also explored the neighboring Italian villages on a motorcycle, met Italian friends, and noticed that they didn't want to harm him.
- When going out after the vocational school blocks in Bolzano, Sigmar also comes into contact with young people from abroad, essentially has good experiences that outweigh some of his friends' negative experiences.
- Sigmar has had an Italian girlfriend for a year; he believes he can be with her longer.

For Sigmar, the connections, especially with the Italian youth in the neighboring Italian villages, are, on the one hand, (trans-)cultural learning moments in conscious reflection; on the other hand, they are one of the reasons why he distances himself from the riflemen. He cannot understand that riflemen's friends from other villages give him a strange look because he also hangs out with Italians. His intercultural experiences, gained with the motorcycle and the motorized tricycle ("Ape") while driving to the neighboring villages, come against a previously unreflectingly approved ethnocentric culture. This irritation probably enables Sigmar to tap into a resource that—unconsciously—he must have had from an early age. Connections with Castello and Capriana correspond to that world of cultural coexistence in a small space in the place of origin of his father, the little Neumarkt faction Laag in the lowlands. Sigmar had spent his first years there before the family moved to Altrei. His father was a riflemen captain in Laag and represented an ethnocentric view of the world in a "mixed language" environment. Still, the family and living conditions of Sigmar were undoubtedly "mixed languages"; for example, Uncle Maurizio, who bought him ice cream, indicates a multilingual family background. In the Habermas model, "system" and "lifeworld" are examples of counterparts: In the "system," Father Decarli is committed to Germanism, in the "lifeworld," Italian is a matter of course.[5]

Another speculation lends itself. In the "mixed-language" conditions in Laag, where the German minority was exposed to a specific need to assert and differentiate against natural assimilation tendencies, Father Decarli sought distance from the other culture and split. In Laag ultimately inevitable—Italian parts of his culture in the system of the riflemen. He moved to his wife's largely undisputed German hometown with the family, which had strong roots there. The attempt to become a farmer can be interpreted as a search for a culturally unbroken

---

[5] The concept of system and lifeworld (Habermas 1984, 1987) is presented here only to the extent necessary for the subsequent case studies. It is explicitly taken up and discussed again in the concluding chapter.

Heimat with secure roots. What was difficult for the father in Laag, namely to venture into the intercultural opening, was achieved by the son in Altrei with far less effort: he was able to launch curiously out of a culturally unchallenged German area to the Italian Val di Fiemme and to make uninhibited friendships there, which for the father, who had the intercultural learning opportunities on his doorstep, was difficult. It is an example of the paradox that Siegfried Baur (2000, p. 171) calls the "pitfalls of proximity," a "proximity with the simultaneous devaluation of proximity." The devaluation of "closeness" serves to construct cultural uniqueness, which offers more security. The relative distance to the other culture provided by a culturally protected microcosm like Altrei, on the other hand, made it easier for the son not only to seek contact actively but to accept it as enriching and—that is the point—to be aware of this in his reflections about the "system" of *Heimat*. It could also be imagined differently: The same Sigmar, the same ultimately protected world, the same excellent relationship with Uncle Maurizio, the same Italian friends—but militant patriotism, whenever the conversation gets from the real-world level to that of the political system.

Sigmar's distance from traditional positions and identity constructions clearly shows personality growth. He places his need to get on well with everyone above every political feeling of belonging, gives the Italian an explicit value, and attaches a positive meaning that he knows many foreigners whose nationalities he proudly lists. His orientation only wavered slightly in the few places where his mother took part in the conversation. On this occasions, he changed from the subject of "oppressed foreigners" to the suppression of the South Tyrolean eagle on T-shirts by teachers and schools and emphasized the need for South Tyrol to remain German, which for him had previously been out of the question, and to such an extent that the actual problem was the waning knowledge of Italian by the Germans. In case of doubt, he also insists that he is "not an Italian," but it bothers him that this might harm well-meaning Italians.

Sigmar likely had a safe childhood. The family home seems intact; the lifeworld in Laag was highly politicized but peaceful within the family. A child has died. Information about this is missing. It was mentioned in 1997/1998 on Sigmar and his mother. I didn't question about which child came in which order at that time, possibly because of my own bias: Norbert, my mother's second child, also died as an infant; it would have been my oldest brother; I tried to clarify this fact and reflect on what it could mean for my family and me only later on. Sigmar's deceased sibling was not mentioned in 2009. But it may have cast a shadow over Sigmar's perfect world, which may have been expressed in one of the threats that 11-year-old Sigmar saw emerging. At the same time, Sigmar could feel particularly connected to the mother, as support for the mother and, for a long time, also as a representative of the deceased child. At a certain distance, Sigmar was succeeded by three siblings, the youngest again a boy who takes after his father—a hunter like him, a radical like him.

Sigmar's bond with his father is undoubtedly strong. The "Tata," he had said in 1997, brought him to the riflemen. However, in his development, he was probably more oriented towards his mother, resulting from her assessment that the younger one takes after the father. In the culturally unequivocal family history on his mother's side, Sigmar may have found the necessary support to let go of the feelings of threat that had been instilled in him as a young rifleman. Despite his affection for his father, he seems to identify more with the conciliatory parts of his mother. Her world seems more solid: She comes from a firmly rooted and recognized family in Altrei, grew up in the traditional world of riflemen, but has a critical attitude towards exaggerations: in 1997, she protested against the notion that Altrei was only an old village with older people. In 2009, her youngest son, Sigmar's brother, is one of the riflemen who carry the controversial crown of thorns during the state parade.[6] The mother can approve of this because the object of the dispute, defused with roses, was only played up politically, and people liked it during the parade.

On the other hand, she rejects the refusal to pay a salute as a protest against politicians who would do too little for self-determination. In his distancing from the riflemen, Sigmar goes further than the mother: the refusal to pay a salute is "disrespectful," unlike his brother, he would never have worn the crown of thorns. But he talks with greater ease, more loudly and freely when his positions match the mother. This could indicate that he owes the easing in his own identity designs more to the mother and therefore wants to stay as close as possible to her, not wanting to cause pain or inconsistency through positions that differ too much. His hesitation when delimits his own positions also shows that developing one's own opinion to free oneself from group pressure and tradition takes labor and effort.

From 1997 to 2009, Sigmar's world changed. Altrei is primarily protected, although not free from the changes caused by modernization and globalization. The road to Altrei, which Sigmar drives every day on the way to work, is broad. Commercial zones tear open the idyll of a mountain landscape. The father finally had to give up his attempt to become a farmer and take up a job as a wage

---

[6] Every 25 years, Innsbruck hosts the commemorative ceremony of the *Land Tirol* (the Austrian Bundesland Tirol and the Italian province South Tyrol) in memory of the 1809 battles, of which Andreas Hofer is the symbol. Since 1959, riflemen have carried a colossal steel crown of thorns in the parade. The comparison of the division of Tyrol with Jesus' suffering on the cross provokes polemics on every occasion.

earner, as a "man for everything"—a *"tuttofare."* Sigmar was unable to fulfill his childhood wish and, as a mechanic, is experiencing how little value his work has—less than sitting in the toll booth at the motorway exits and collecting the toll. Where the mother perceives one or the other threat (the EU, the foreign people who invade the forests), Sigmar remains consistently conciliatory, softens, relativizes. He is focused on the future, earning enough, and building a house with his girlfriend, not an apartment, but a real house. The fact that she is studying encourages him that once she has a doctorate, she will earn more; with that, they can make it. In Sigmar's world, too, strangers and strangeness come in, but somehow he gets along with the fact that he is interested in the stranger, that he wants to start the conversation, that he "tries" to get along well with everyone. He has kept an outlet for possible aggression, fears, and images of threats: Italian politics, which he connects to Silvio Berlusconi, but also describes with a strange-sounding formula "below, above," below in Italy, above in power.

Speaking from Habermas: Where changes problematized Sigmar's lifeworld, he was able to take advantage of opportunities for communicative action and suppress the colonization of the lifeworld through the "politics" system. Sigmar's lifeworld, which is communicatively and emotionally accessible, was thereby expanded and freed mainly from the phantasm of the threat that would be inherent in the *"Heimat* system." As a result, he was able to dare encounter beyond the cultural borders. These openings allow him also enter into an interethnic partnership with an Italian girlfriend with whom he—with a critical eye on the dark side of his life (the barren social prospects, the alienation of an ideal world)—wants to spend his life.

Ingo Hört

But You Don't Need a Traditional Costume for That

Ingo Hört, Prad, Riflemen Company Prad, 14 years old; Interview on August 30, 1997, in Goldrain, in-depth discussion in his parents' house in Prad on November 29, 1997.

> *"When I am in costume, then … then I am there to represent this country … that I am here for this country or something …*

What is this country for you?

Huh?

Is this country that you want to be there for the area, the nature, or the idea that it will remain German?

It is nature, and the fact that it remains German also matters, the language ... the tradition should always stay the same.

Is that good? Shouldn't things also develop?

Yes, they should develop, but always develop for the better and not change everything. That shouldn't happen. It should be more modern, keep up with the times, and not go backward."

"Italians ... we have bad and good ones here. It is the same for the Italians. [...] Yes, there are also Italians here in Prad, just of Italian descent or who just ... live there."

"What kind of music do you like to listen to most?

Hard rock. [...].

Böhse Onkelz?

No, that's kind of too nationalistic to me... or they used to be... but they're still nationalistic and somehow don't want to admit it. I think that's kind of cowardly."

At the campground in Goldrain, Ingo wears a T-shirt from the hard rock group Slaver with a steel-claw-reinforced eagle on the front. He presents himself as cool, grumpily drops his long strand of hair on his face, has his hands in his pocket, but is not abrasive. For a picture, he willingly takes out the rifleman hat he brought in a nylon bag, puts it on, turns and turns as desired, and according to the lighting. In contrast, he stands side by side next to his friend, who has a red Tyrolean eagle emblazoned on his back.

He replied sparingly but thoughtfully to all questions, sometimes leaving it with short assessments ("oh, you know"). He didn't want to be nationalistic. To him, *Heimat* is of a "certain nature," then he falls directly into the more common response pattern and thinks it is about "representing" the country. When asked whether he meant to defend, he insisted: "represent." He adds a bit indistinctly: "... because everyone who lives for their ... *ation* should represent it." When asked whether he had said "tradition," he replied: "Yes, tradition, nation."

Ingo only likes hard rock, Slaver above all. He rejects the cult band "Böhse Onkelz," which started with radical right-wing songs and then became milder. His answers are hesitant but accurate and considerate. Ingo came to the riflemen because of his grandfather, but that doesn't mean that the riflemen were old-fashioned. It was only "bad" if they were "nationalistic" or "so right-wing radical." He does not like nationalism. He also mistrusts the alleged conversion of the "Böhse Onkelz"—Ingo sometimes has a somewhat suspicious look, but it probably only expresses thoughtfulness.

The next day, at the fair, he has to pull himself together again and again. Ingo is next to the state commander and self-determination politician, Eva Klotz. He holds the flag of his district, sometimes yawns.

Telephone conversation with the mother. The same sentences for the umpteenth time: exhibition in Hall in Tyrol, with which the exhibition hall Tyrol is inaugurated, feelings of *Heimat* among young riflemen and marketers, interviewed their son at the tent camp in Goldrain, would like to come. "Ingo will be there," declares the mother, and does not consider it necessary to make an appointment. November 29, 1997: Ingo is here. He smiles a little embarrassed, has cut his hair, brushed it up, and dyed it.

A few warm-up questions: Ingo answers again thoughtfully, earnestly trying to report his, not any opinion. Only casually, on a random question, does he mention that he is no longer with the riflemen: "I have now quit, also because of Goldrain ... the Nazis there got on my nerves and otherwise ... I have changed my mind."

Ingo now declares himself a pacifist. Since he started doing this, he doesn't like going to the youth room anymore. He doesn't want to say why. Shrugs. A few days later, the story is on TV. A small group in the youth club defines itself as "Nazis"; Swastikas are painted on the abandoned train station and school. Non-Nazis are mobbed, occasionally beaten up. On the washed-out station walls, they sprayed their answers next to the swastikas: "Nazis piss off," "stop animal testing, take Nazis." The "Nazis" have been contemptuously called "wanna-be Nazis" in the youth club since they became defensive due to the public scandal.

A visit to the youth room is like opening a book on youth culture. The group is self-governing. The rule is that there is no hitting in the club itself. It is the non-violent room in the village where the camps meet without going at each other. Everyone takes their beer and their girl and goes into a corner. Only a few derogatory glances wander back and forth, sometimes a wow, a raised middle finger: "I wank you to death" is one of the sayings at the train station. Manuel, the leader of the so-called wanna-be Nazis, defines himself "as not quite a Nazi, more like a Skinhead, a bit stupid, nothing more." What exactly does he mean by this? "Beating someone up when they get on our balls, I mean beating them up but not quite ready for the hospital." Manuel opened the skinhead cult at a holiday camp in Germany. He was there when they "beat up a black man," he liked that. He doesn't know precisely what is on the right. He has heard of concentration camps before but has not dealt with them. Boh. They wear spring boots and secretly smoke marijuana. Have their own fields. The police found some, but not their fields. They grin. They hate foreigners because they steal. They have broken in somewhere themselves, someone admits.

Prad is a finely cleaned village, the first on the exciting and winding road to the Stilfserjoch, the Passo Stelvio, the highest pass in the Western Alps. But the village lays still on the valley floor, not far from the main street. From nearby Ortler and the Sulden glacier ski area, it has transit tourism, the epithet "on the Stilfserjoch," and the honorary citizen Gustav Thöni, through his Olympic and World Cup victories in skiing and radiant modesty a kind of model for the South Tyrolean in the early autonomy period. In many rural communities, exclusively Germans are represented in the municipal council. In the latest census, only 94 of 3027 citizens declared as members of the Italian language group. Most of them came here as Carabinieri or financiers, have German wives, mostly speak moderate to good German and send their children to German schools. The Italian primary school was abandoned a few years ago due to the lack of children. There are practically no foreigners, except those who work in the hotels. Unemployment is not an issue. Prad lives primarily from an up-and-coming small and medium-sized industry. The most important employer is a door handle factory in Schluderns outside on the main street.

Death keeps making headlines for Prad. In the youth club, there is a photo of Christian, who died on November 19, 1997, on the road. The succinct aspect of the ad expresses the repetitive nature of such accidents. There is only one tavern ("Ladum"); the youngsters travel from village to village at night to find places where something is happening after the traditional curfew. Alcohol consumption and racing with the Carabinieri are part of the youth culture. A collective suicide case brought the village into the headlines across Italy: four youths committed suicide together.

Another suicide case was taboo for a long time and is still controversial. In the South Tyrolean bombing years, rifle captain Josef Tschenett was arrested as a suspected Tyrolean terrorist. Two years later, the man was still in prison, and his wife was found dead in the river Suldenbach. Three of their four children were still underage. The village book only says: "In the difficult political years at that time (1961/1962), Josef Tschenett resigned as a captain." (Loose 1997, p. 387). Not a word, neither a reference to the attacks and the arrest nor the death of his wife. Some of the children of Tschenett suspect that the mother did not kill herself but was killed (Tschenett 2004).[7] Tschenett's successor in the years of the assassination was Eduard Fahrner, Ingo's grandfather.

---

[7] One of the sons, Stefan Tschenett, has literarily captured the history of his parents and especially the mother's death in the novel Feuernachtsmord (*Fire Night Murder*).

Ingo has two brothers; Freddy is 20, Markus is 24. Both were with the riflemen before him. They been out for a long time but would have wanted him to stay because of the tradition "passed on from grandfather." In Prad, the riflemen were not nationalistic; that was not Ingo's problem. But in Meran, he knows a young rifleman who has an old German flag in his room on one side and the Tyrolean coat of arms on the other side. When "the Nazis" appeared in Goldrain, he had enough. He recognizes them "by the white shoelaces in the jumpers ... and by the bald heads". It was simply no longer his world. The mother says, "It's a pity." Ingo scratches his hair: "It's true that one should stand up for the *Heimat*... but not against foreigners... not with demarcation... after all, everyone comes together. Blacks and whites have to live together." The Nazis in the youth center would say about the foreigners that they would take their work away from them. Still, he knows that this is not true.

Ingo spent the last year of secondary school in a home in the Johanneum in Meran. He wanted to get out of the village. A friend had gone there the year before. Now he is attending the language school in Schlanders, the first class, commuting. He likes to read, and he is sure to be good at German and would like to "work for the newspaper... write articles".

The Hörts live in a new building outside the village, in one of the new settlements, in which almost all houses have the same number, pretty, neat, but not yet familiar *Heimat*. When asked what his closest *Heimat* is, Ingo says firmly: "No, not the house." The room? "That yes." The chamber says a lot about Ingo. He stands in the middle, points around a bit, sits on the bed, lets the room tell about himself, a Palestinian sash wrapped around the lampshade, posters, a skateboard. His band is no longer Slaver but Sepultura, a Brazilian hard rock band. Ingo's brothers are already playing in a band. He wants to start one now. Michael Jordan, the basketball star. A role model? "Yes, in a way." Ingo already has five people for a basketball team. He would need to use the middle school gym. The skateboard? "Yes, he is doing that too now." A picture of Che Guevara. Is he a role model? "No, if he had won, he would have become like Fidel Castro." When asked whether he combined the Palestinian scarf with a political attitude, he says, "This is a cloak of the Jews or something."

Ingo looks around: "Yes, the room is a bit of a contradiction ... it looks somehow ... there is not much Tyrol, actually ... nothing at all." No posters were stuck in the room as long as he was with the riflemen. The cans were there before, and the radio wasn't. But he already had the idea for the band when he was still with the riflemen. He returned the costume. He is now wearing wide sweaters. "Faith was more important before," he says. Now he feels "freer." He found being with the riflemen "exhausting." Always alert. Always upright. Always one hundred percent. "Yes, in a way." And what was nice about being with the riflemen? "That you are together in the community. The emotions. *Heimat*. Have a hold. Pride too, honor." All of this was still important to him, but "everything was a bit more fanatic by the riflemen." "Somehow," says Ingo, he wants to "still be interested in Tyrol. But this does not include a traditional costume, it does not mean that you are a rifleman ... and that you have to prove your honor".

Identity as a Creative Game with the Risk of Being

Summary description 2010; Interview on October 14, 2009, in a rented apartment in Meran, where Ingo lives with his girlfriend (absent); he is 26 years old, lived and worked in Vienna for a while, but then moved back to South Tyrol. He immediately takes up his statement from 1997 that his actual *Heimat* is his room. Nothing has changed there, even if he has changed rooms frequently in the meantime; he feels at home where he can retreat.

> *"Because when you go out on the street, you can never really do anything... you can be yourself, but then people look at you strangely... I probably had problems with that... I still have that today that I often think about what people think of me. Or if someone looks at me now, what is bothering them about me ..."*

> *"A colleague of mine had a pretty ... a pretty bad car accident. And actually, from that moment on, I thought it was really better if I was at home, because if things like that happen, especially in the family, like if my brother somehow has a car accident and ... his life is in danger ... or something. It would take me nine hours to be there, and I could never do that. I would never forgive myself if I weren't there."*

> *"Yes, I am probably not talking about what you expect ... But I have absolutely no desire to talk about South Tyrol or Italy or anything ... The subject is so ... I don't know ... that's too much for me somehow, and it is to me ... to me, it is actually really unimportant. It does not matter to me where there is a border somewhere and where there is not. I could never say that South Tyrol is my Heimat because I know about two percent of Puster Valley, maybe I saw that when I was a little child ... intanto ... I know the area between Landeck and Vienna better than the Ulten Valley."*

> *"What do I care about my ID? If I didn't have to have it, I would throw it away."*

> *"Yes, identity is completely individual ...".*

> *"Actually, you can say that I am the Heimat for myself."*

"Yes, I still remember how I came back from Vienna and saw the mountains, then I did feel ... I somehow feel it down my spine ..."

"I do like the traditional costumes, I have to admit that ... or if I see 200 riflemen or more somehow ... marching, then it'll raise my hair, that's something that means something to me."

Contacting Ingo is easy in 2009, he has a Facebook page on which he is actively present, maintains a straightforward, mostly dialect small talk with a growing "circle of friends" (as of August 2010: 456 Facebook friends) and an extravagant appearance: In 2009, his profile picture is just a snippet of chin, mouth, and nose, processed and alienated in Photoshop, in August 2010 he wears a Superman emblem on his chest and fires rays from his eyes. He replies to an email asking for an interview quickly and easily: "Glad to hear from you" I should just come to the *Kunsthaus Café* in the evening.

We first meet in the Café in Meran, a stylish eatery under the arcades, which offers a stimulating cultural program and is—faithful to the name, in English Arthouse—a reference point for the local art scene. Ingo works as a waiter here. As of August 2010, it is his last professional stop. He has all sorts of things behind him: he started school but dropped out again because his father thought it was enough after the second fail in three school years. So he started working at Polyfaser in Prad, a swimming pool manufacturer, where his father and mother also made their money. But he could not endure fiddling with synthetic resins and glass fibers. A colleague from Prad, who studied prehistory and early history in Vienna, placed him in Vienna with a job in archaeological excavations. Ingo liked that. He got to know people from all sorts of countries, felt comfortable. After returning from Vienna, he started working again at Polyfaser until he found a job in a retirement home. He liked the job, but he quit due to internal management problems. He was unemployed for six months. Then he had the opportunity to work as a computer tutor in the Tangram training organization in Meran—one of his passions; Ingo also mixes his music on the computer and processes his artistic graphics and photos. When the project contract expired, he was out of work again, felt "annoyed" by everyone who wanted to push him to new work, enjoyed free life: "You just have time for things that you can never do otherwise." And then the head of the *Kunsthaus* reached out to a cult pub in Meran with Italian beer, which to Ingo tastes better than any South Tyrolean or German brew.

After work, we go to Ingo's home. The rooms are minimalistic, empty, with little furniture, an extravagant style. Since 1998, when his room swelled with symbolism, some of which he could hardly assign, Ingo has reduced his aesthetics to an extreme; his hair is short-cropped. What is still reminiscent of the young rifleman who clings to the standard is perhaps the shyness in his eyes. And there is the early confession in the first sentences that the question of whether others are watching him, whether they let him be who he is, or whether they look at him strangely is an important one for him. He feels secure in his room, where he knows to be unobserved. At the same time, it is striking that he sometimes gauges whether he is saying what is desired or whether he drifts too far away from the subject of "*Heimat.*"

Ingo feels his way forward in a world that he experiences on the one hand as threatening and, on the other hand, as tempting. The threat comes not from the vastness or the foreign but the close and near. It was not in Vienna or when dealing with people from everywhere that he asked himself what the others wanted. But if he goes down from his apartment in the Meran arbors in the morning and is being looked at strangely, then he thinks that if he weren't from here if he was maybe even a black man, he wouldn't feel comfortable being looked at like that; it might even scare him. At the same time, he couldn't stand the idea in Vienna that something could happen to good friends at home or to grandma or his brothers and that it would take eight or nine hours for him to be with them. The severe accident of a friend, who fortunately survived, triggered the decision to give up work in Vienna and go to South Tyrol—despite the tightness that he can hardly stand, but because of the few people whose proximity he allows.

The familiar and the foreign play a surprising game with Ingo. It does not seem that he has to project and dispose of unconscious foreign parts in fear images, but the opposite: as if he would project the uncomfortable aspects of his culture onto fear and threat images and find protection in those cultural codes that are foreign to him, in wild music, in distorted photo design, in punk culture, even if he ultimately weighs the outward appearance just as skeptically as that of his tradition.

After leaving the riflemen, he had a punk as a friend and liked to irritate his father with colored hair and run around with dog chains on his clothes. He kept some punk ideology, but it became too bland and formal for him, almost like a costume or a uniform. At the same time, he admits that he also depends on the symbolism of his "own" culture: the mountains, the rifle marches, and the Sacred Heart fire with their beautiful moments in memory. In the annual Tyrolean Sacred Heart celebration, spiritual mysticism and political myth meet the custom of lighting fires on the mountains on the third Sunday after Whitsun originates from the solstice celebrations. The tradition was reinterpreted in 1796 when the Tyroleans called the Sacred Heart of Jesus to protect the *Land* from the advancing Napoleonic troops and, in return, was promised eternal faithfulness. The climax of the South Tyrolean attacks in the 1960s, the *Feuernacht* (night of fire), took place on Sacred Heart Sunday in 1961.

Ingo comes from a patriotic and tradition-conscious family on the mother's side. Grandpa was the riflemen captain in the bombing years. It was an occasion for him to join the riflemen, and it was one of the few reasons he regretted leaving in 1997/1998. Even his brothers, who had gone before him, one six and the other ten years older than Ingo, regretted this break with tradition.

What he had not mentioned in 1997, he reluctantly tells in 2009, although without shame, but somewhat unsure whether it is appropriate: "I have to say something else ... May I?" Ingo is his father's only biological son; the two older brothers come from two other fathers. Even if the father had never drawn a distinction and both brothers would say "Tata" to his father, Ingo still believes that this could make him unconsciously more confident of finding unbreakable support in the father even in the event of difficult conflicts. When he first came home with green hair, the father said, "You look like a gypsy. Go up to your room". The father didn't speak to him for two weeks, but when Ingo still hadn't cut and bleached the hair, the father accepted it. Ingo does not take it for granted that he could survive the conflict. "My Tata wasn't brutal, my Tata didn't go and hit me like others ... many do, I think. That may be a difficult hurdle that many simply do not take because the parents are too brutal." The fact that his brothers did "a lot more... things with the police or otherwise" made it easier for him to be different. He was lucky not to get caught.

Ingo would have known in 1997 that he belonged to a patchwork family. The mother told him so when he was eleven or twelve years old. In a family that endures the different origins of the children and their transgressions, something like a basic trust should develop—despite all the difficulties—that even deviations from the norm are ultimately absorbed with love, that even failure does not mean exclusion from the family. Ingo should have noticed from an early age that "one"—within the family—could also be different, could come from other fathers, could also get in trouble, with paternal anger and maternal concern, but without risking the belonging to the family. This could be an encouragement that enabled him to leave the riflemen, where his grandfather was highly honored, and walk through the village with dyed hair, even though his father did not tolerate it. A patchwork family that does not hide its cracks, but holds together well, could have encouraged Ingo to be different. At the same time, his parents' home was also well integrated into the South Tyrolean culture. When Ingo leaves the riflemen, he can also create a distance to the parental world of values, trusting that he has a hold in the family home that will survive this break.

But this step did mean: be a stranger in your own ambiance, in your own village. After leaving the riflemen, he avoided the youth club, where the "Nazis" or "wannabe Nazis" had created a climate he did not like. One criterion according to which Ingo defines his affiliations is well-being. When confronted with the radicals, he no longer felt comfortable with the riflemen. Riflemen friends, whom he knew well before his time as a rifleman, remained with him. The others fell away. The change of location first by visiting home, then through high school, and finally, his time in Vienna opened his world a little each time. Each of these steps out of the narrowest microcosm was facilitated by the fact that good friends had gone ahead—to the Johanneum, to the subculture of the punks, to Vienna. Concern about a friend who had an accident leads him to return home. It is part of the history of Prad that young people suddenly die away—the multiple suicides not yet forgotten in Prad, but also repeated deaths on the streets, mostly late at night.

Seen in this way, Ingo's *Heimat* is above all a close relationship with his family and his few friends, which—as he emphasizes—no longer increases despite the ever-growing circle of "colleagues." Grounded in this security, he moves, dares extravagant looks with his exterior, tries out styles, endures being unemployed without losing his nerves, but on the contrary, enjoys that he could be more creative during this time. The only thing that annoyed him about unemployment was the concern of close people when they were worried about him. Those close to Ingo are important to him, he gets courage and strength from them, but this high value of solidarity is also the only thing that limits him, what brings him back, what binds him to a place. Otherwise, he says, his *Heimat* is the whole world, and even this would be too narrow if he could go beyond it.

Little is fixed with Ingo: he has freed himself from tradition, interprets faith freely, and has no training that would point in one direction. He had taught himself a lot of what he had learned: English, later Italian, music, and computer skills. Ultimately, he trusts that he will find something that suits him. He leaves behind what doesn't suit him: his parents' "shit work," school, when his father said that now it was enough, work in the old people's home because of the problems with leadership. Ingo seems to have a reconciled, clear relationship with his paternal authority. After time had passed, he spoke to his father about the two-week ban in response to the green hair. Ingo thinks that the father couldn't remember entirely but at least could address it. With foreign paternal authority, such as a boss, he does not mind having problems but does not look for them. As long as it fits, he is willing to compromise, and if it goes too far, he stops. Ingo, it seems, is more likely to avoid authority conflicts, but it could also mean that he does not have to pick substitute fights.

Ingo's *Heimat* or identity design is postmodern. Without clarity, it exposes itself to a stream of styles and codices and plays with them. What could make you wonder about the needed withdrawal to his room at home based on the sensation of being looked at by people in a strange way? Is a draft identity that foregoes the security of clear cultural affiliation associated with the price of such disquiet? Ingo risks being the way he wants to be. Even after work, he likes to go to his pub with his provocative skullcap, where he then usually adapts with his clothes not to scare customers. But he also feels that this provokes a fear of persecution in him, namely: how others look at him. The idea of what it would be like if he weren't a local gives him naked, existential fear. That he nevertheless risks it, provokes it, also means: wanting to feel a little bit strange because what is one's own could be too suffocating. In South Tyrol, everything "stands": This is not a phallic metaphor for Ingo, but borrowed from the image of standing water in which nothing flows anymore because life begins to rot, suffocate. If a body of water stands too long, life gags inside it.

Ingo creates freedom at the price of the feeling of being viewed unpleasantly by others. Abstracted from Ingo's situation, this could be seen as the flip side of that delusional notion that one's own is threatened by others. Here it is the stranger that Ingo identifies with and is threatened by one's own. It is not the falling apart of his identity that frightens him, but that he is forced into a unifying identity, to the extent that he ultimately sees a uniform in the shimmering surface of punks. What is admirable about punks is not that they are all the same, but that they are all different: Only this creates the freedom that Ingo needs.

Perhaps one can now understand that he places so much value on his family and friendships: exposing oneself to the flow of the non-conforming, the diverging, means to go without the reflection and confirmation by one's cultural affiliation; it is all the more important to him to keep the connection to the family and his friends strong, not to be too far away if something could happen to them. Vice versa, it is also unconsciously thought that they are not too far away if something could happen to them. The firm anchoring in a miniature cosmos of family and friendship, in which diversity has been and will be endured, allows Ingo to consider the whole world as *Heimat* and be curious about other languages, people from different origins, about styles that clash with cultural norms of his heritage. An intensely experienced and cultivated lifeworld disempowers the system because although it threatens to protrude into the lifeworld, it encounters strong bonds there that can hardly harm it.

Ambivalences remain, but they do not disturb. They are part of it: advertising for tourism, the label that a country gets bothers him because of the outdated imagery; if it were glaring, surprising, groovy, he could better identify with it.

This expresses a wish that his *Heimat* alienates himself to get closer to him, which has made him alien. For Ingo, identity is "fully individual." One after the other, he rejects all possible criteria: Belonging to a "country" creates no identity for him. He has relativized the meaning of the language since he knows English and more Italian, and identity about work is hardly an option for him. As if he had to create security somewhere, he suddenly comes to speak of his "roots" and locates them in Christianity. By thinking that "90 percent are Christians", he suddenly conceives *Heimat* as the majority's culture but immediately puts the idea into perspective. He doesn't stick to it, and he frees himself. Ingo's thinking, for example, shows how strongly *Heimat* and identity designs are related to the need for security through external- and self-assurance. Where there is no external assurance, there is a bit of uncertainty, which of course, also offers scope for your own design.

For Ingo, "the coolest" is "doing something creative with friends." He performs small gigs with a friend and a guitar. He experiments with music videos with another friend who is an artist. Once, this friend exhibited a picture of Ingo with his girlfriend: on her bare torso, she wears a duck head, in one hand she is swinging a meat cleaver, with the other she holds Ingo's head as if she were about to cut his neck. Ingo clicks his tongue when he describes the picture: "... and the people very well received it... it looks terrific, it is a perfect photo, and everyone knew me, and everyone asked me about it, ah, that's you, and back and forth, and I find things like that so neat. We did that together." For Ingo, exposing yourself to the horror of human existence playfully and creatively means having to shudder a little less in your own world. Or: From a secure friendship, you can also play with someone's head being cut off at any time.

Postscript
Ingo changed his Facebook profile again in March 2011; now, he is holding a newborn, his son, and we continued to keep in touch by FB. A glance at his Facebook page in July 2020 shows him standing under a cone of light with folded hands—experimenting remains an open game. He is looking for an apartment for a dear friend in a posting. In another one, June 16, 2020, he makes it clear: "I'm fed up with racist posts. The next one, who thinks he can judge people by skin color, religious beliefs, or origin, is removed from my list of friends. 2020. People, wake up."

Dagmar Lafogler

Just Like Andreas Hofer, Sure, I Would Risk that for my Country

Dagmar Lafogler, Rabland, Schlanders, company Schlanders, 14 years old; Interview on August 30, 1997, in Goldrain, in-depth conversation during the journey from the school in St. Pauls to the mother's house in Rabland and the garden.

"I also find it really unfair that, for example, the Italians don't have to learn German, but we have to learn Italian. [...] The Italians do not accept it. The German is not accepted. You saw that in Bolzano when they broke the Franz Innerhofer plaque. Italians don't accept our culture as we think, only theirs. If you did something like that to them, you'd be jailed for a day."

"Well, I think the Italians have no right to downgrade our culture to say that we only drink, that we are only there to drink, and other things."

"Yes, and I think it's unfair because you can do what you want in the south. Well, then they should stay south, and when they come up, they should accept what kind of culture we have. Because frankly, I am often ashamed that South Tyrol belongs to Italy."

"I can also use violence when it gets too much. [...] One time it went like this when two came and looked at us ... [...] it was the riflemen, they were drunkards and things like that ... and when someone started grabbing me and stuff ... and saying, for a rifleman you put up with a lot, because I know that you marketers are all sluts who let everyone get some, then it was done. Then I just demolished him because I don't care whether it's a dude who is three times as strong as me or who is half of me. I don't give a shit. It's easy. If I start scratching right now and really feel the thing that I hate, then nothing can stop me ... [...] He's got his nose cut off. I have a smaller eye for it."

"Well, Andreas Hofer is my role model ... I am thrilled at how he defended Tyrol, and I am somehow proud that I am with the riflemen, and I am not ashamed because I don't care what other people say, they can talk until they drop dead. [...] I think... if it is an extreme case where everyone is really intimidated or something ... when it is really so far that everyone is intimidated, it is too late. Surely then someone has to pull everyone along like Andreas Hofer, for example ... he pulled his people along. So I imagine that I would also pull my people along, and if it would cost my life, sure, I would risk it for my country."

Dagmar is a hiking child from Vinschgau, in Italian the Venosta Valley. She lives in Rabland, one of the first villages from Meran via the Töll into the Vinschgau. As a rifleman, she is at home in Schlanders, the valley's main town, where her mother has also been for 15 years. She grew up in Naturns, the most urban village in the Lower Vinschgau region, where the hotels and the town hall are fashionable for tourists. The wind is sharper here than anywhere else in the country, and it is said that it carries people and souls far.

The villages where Dagmar was and is somehow at home are on an old Roman road. Rabland was originally called Cutraun or Catraun after a Roman-ized Celtic name. As the later Germanized "Kaltraun" finally became Rabland in Germany, the priest Josef Ladurner (1770–1832) caused a mischievous reckoning with his colorful people: "Vagrants, beggar, cart puller, troops, traveling sales-men, Raven people. This made Kaltraun to Rabeland and took away its honor and good name." (Lassnig 2012, p. 420) The *Korrner*, a Tyrolean gypsy variant, equally ostracized, equally persecuted, drove into the jet-black hair and ease of the ancestors out of revenge.

Between Naturns, where Dagmar grew up, and Rabland, where she lives, there was once an angry love dispute that took on the proportions of a little Troy by Tyrolean standards. A young knight named Gerold had married the young farmer's daughter of the Steinerhaus, today's inn "Hanswirt," not quite appro-priately. His uncle Kuno von Hochnaturns was so angry about the shame that after Gerold's death, he locked the widow in the "rave tower." In a considerable uprising, the peasants freed their own from the tower. As if the history of a place leaves traces in people's lives, an invisible line leads from Rabland to St. Pauls in Eppan. The valuable Margaret statue in the old Jakobus church in Rabland is said to have belonged to a single altar with the figure of St. Jacob in the parish church of St. Paul.

Dagmar attended the last year of middle school in St. Pauls. Far away from home to focus and finish the final school year. Since her parents' separation, she no longer has a connection with Naturns. Only her boyfriend lives there. She thinks Rabland is "quite okay," but there is no meeting place for the youth in her village so that people "don't just have to hang around." Most of the Vinschgau villages, but especially Rabland and Naturns, have become street villages, one of which has been robbed by the draft of the main road. It goes either into the valley or out, down to Meran and further. Since the four-lane road from Bozen to Meran has been expanded, the trucks have been rolling in columns at rush hour through the villages, towards Nauders in Austria, or through Münstertal to Switzerland.

Dagmar would love to do the interview at the college dorm in St. Pauls. A little homelessness seems to sound through the phone as she says it, or even the caution not to look very closely into your world. The sisters of the monastery school would have allowed her to get her own room but didn't like it in Rabland.

Then she raves that photos can be taken in the garden behind her house, it is beautiful there, in winter, especially when the sun glistens in the snow, but she does not know whether the mother agrees. Finally, she agrees that we pick her up at the college dorm in St. Pauls on a Saturday and drive from there to Rabland. It is now spring.

She comes shining across the schoolyard, puts her bags in the car, and slides in a cassette. At the youth shooters' camp in Goldrain, she had confessed to the "Böhse Onkelz" ("this is my group"), even if these are forbidden: "I like the texts because they write texts with a good meaning, and then I feel that no matter what the others say, just because I hear the Böhse Onkelz and I am with the riflemen, it is connected to the Nazis again and then it is connected to the alcohol and then it is connected to the drugs and then again to the sex and all that stuff." Whether she still likes the same music? Yes, she laughs, but in the meantime, she also hears something else, "Natalie Imbruglia and everything else that is modern, including English." What she doesn't like is "this folk music, corny music."

Dagmar is 14, and her boyfriend is 19. Coffee break in a restaurant that is about to close. She speaks across the empty tables without fear. She already has clear ideas about life. This year she will finish middle school, she says, fingering a Marlboro out of the package. "I have set myself goals so that I can get order into my life. I need that." Some of her friends had tried drugs, but not her, so she had been excluded from the clique. But she always goes her way. She had to get along with herself... "That was the most important thing." She mentions some guilt, and when asked about the date of birth, September 5, 1983, she casually adds that her mother almost died in the process.

No thanks, no bread roll. She had stuffed herself long enough. She would have to adapt to a new costume, but it was too complicated because she constantly gained and lost weight. But now, she is on the right track. The old costume is too short for her skirt, tight for her chest, but too wide for her belly. She has lost six kilos, she says proudly, but she can no longer go out with such a costume, "I am making myself look ridiculous." She will do an apprenticeship for mechanics, the fellow's examination, and the craftsman's diploma. If things go well financially, she would like to have her own workshop, preferably for motorcycles.

With an A, she would get a 125cc road machine after finishing high school, maybe a Yamaha or a Benelli, but it could also be a motocross machine, preferably the Husqvarna. If she gets a B, the mother only gives her the money for a scooter with a displacement of 50cc. But then she would rather buy a more giant machine with her own money. In summer, she wants to go to St. Anton as a waitress. Her father is a cook; the mother learned to be an interpreter but now does housework for other people.

At the interviews in Goldrain, Dagmar had been the sharpest, integrated herself in a conversation because her friend's statements didn't seem clear enough to her. "*Heimat*," she had said, "is my pride and joy," when asked, she narrowed her home to South Tyrol. North Tyrol was not her "main thing." Then she immediately spoke of what was close to her heart—"the injustices" done to the *Heimat* and the riflemen. She has no problem saying that what the Walschen[8] do to South Tyrol is an injustice. She has no problem saying that she hates the foreigners when they take their places away from the South Tyroleans, as is constantly happening to a friend. Andreas Hofer was her role model because he had brought the people along in the fight against injustice, and she would do that too, and if she had to give up her life, "yes, sure, I would do that."

At the Sunday mass the following day, she stood defiantly and tightly, blinking in the sun as if to fear an enemy. Now she laughs, shines, is frank, bubbly, even if she still says the same sentences: "I hate the foreigners; really, I hate them," she says and laughs. In Schlanders, where she was born and is still a member of the riflemen company, refugees were quartered in the barracks after the Yugoslavian war. Most of them have somehow integrated. The children already speak the dialect of the Vinschgau Valley. Some are noticeable, some are not, some are liked, some not. The valley that swallowed the Korner will also cope with the Bosnians.

Dagmar sees it differently: "You know, there are always more, they don't even look for work, but go out on the streets. You have to be afraid at night when you go somewhere that they won't attack you. Because I know ... because I can't stand them anyway, because I mean ... we went for a drink once as usual, then they came, and five of them approached my colleague with a knife. That's the point where I think, oh well, hold on. You are crazy. You know, I don't mean to say that everyone is the same ... sure, there are good people among them. Sure, I know a few who are really hard-working and are trying to get it right to set up a life. There are certainly good people among the Italians or the police. I don't want to say anything about that. But there are far too many bad people where you no longer have confidence."

---

[8] The term "Walschen" refers to the Italians; the expression comes from "Welsch" for members of neighboring Romansh or Romanized Celtic peoples and has initially been neutral in value. It was simply the name of the Italian-speaking part of Tyrol, Welschtirol (see footnote 9), without being meant negatively for centuries. After the annexation of South Tyrol by Italy, the term was given a negative and derogatory connotation for "Italians," although it is often used unreflectively even today.

She wanted to join the riflemen when she was nine and begged for it. Her father was against it, and her mother worried because she was "so radical." At eleven, she finally got her will and became a young marketer. In shooting, she was once district champion and once a national champion. She winds down her shooting performance as routinely as an experienced old rifleman: in Goldrain, it was 283 points in two series, now she recounts what she got: gold twice with a gold wreath, gold once with a silver wreath, gold twice...

Target shooting is a minor aspect of being a rifleman. It's about more than that, she says. What she'd like to be is Captain. To pull the people along like Andreas did, even if it would cost them their life, she could imagine, "..., but unfortunately it is just that us ... us women are being put in our place". But that's also not what matters because there would always be someone for you with the riflemen. "Our company goes through fire for you and sticks together. If you are missing something, you can go there, see that they come to advise you, look to cheer you up. They are just there for you. Everyone puts their hand in the fire for everyone, and that's important. It kind of gives you encouragement and support". She speaks such words with holy seriousness: the feeling that nobody lets anything happen to the other. Her mother's cheerful laugh when she went there, teasing her comrades. The experience that people were friendly to her: "Everyone was nice to me."

Her stories are stories full of scars. During the memorial march for the teacher Franz Innerhofer, who fascists killed in 1921, "they broke the plaque for us." During a vacation in Caorle, Italians were hostile to her because the text "One Tyrol" was emblazoned on her backpack: "They became so mad that I thought they were going to shoot me." A former Italian teacher once pressed her against the wall, slapped her another time, and unjustly sent her home one time with a message saying that she had smoked at school.

All of that because she is with the riflemen. She firmly believes that. Then she pulls her sparkling eyes together as if the sun was dazzling her to tell herself that she can defend herself. It's always the same story. When she was sitting in the bar with colleagues, there were five of them attacking a friend of hers, just like that, she says. That's why she doesn't put up with anything anymore, preferred to hit when a—German-speaking South Tyrolean—guy provoked and groped her because riflemen would let everyone get some. Dagmar broke her bone over her eye at the time, but she did not regret it: "I have defended my country, and I will always defend it, and if I have to pay for it with my life." Eyes closed, tears choked back, hitting for an idea much bigger than her—never be the girl who sobs and sniffles because she doesn't know how to help herself. Behind the laughter, the friendly glow, the squinting look, it sometimes hurts like hell.

She would always know immediately if she could trust someone. She rejected her mother's first boyfriend because she felt how he wanted to get rid of her. When she discovered the pizza restaurant owner who was nice to her, she dragged her mother there until it worked out. Now she says "stepdaddy" when she talks about him. From him, she got Nero, the dog. She just loves animals like crazy, including spiders and snakes. Once she had a dancing mouse, but it died.

Dagmar tells other stories, so the one from a friend abused by their father. First, she couldn't believe it, but it was true. How can you believe something like that? She asks. How should you cope with something like that when you see the father walking through the village all friendly? She talks about a girl who had been waiting for her father in front of the school gate for hours, it would have been his visiting day, but he had forgotten it. The girl just sat there and cried through the afternoon because she didn't want to call the mother to rat out the father. When the father was still at home, the girl once washed the whole staircase so that the mother would not notice that he had vomited on everything. She knew that she was the most important thing to her father, even if it does not always look so easy after a separation ... no, she holds no grudges against him.

Dagmar looks for beautiful places for the photos, an ideal world. The garden behind the house, where the sun shines in winter, the dog, a bench on the way to Sonnenberg, a wall on the street where she lives. She puts on the costume for them, even though it no longer fits her, and then changes into the clothes she prefers: "My jeans and my combat boots, I just feel a lot better in these." The prejudices between the two clothes are at the same time a label: "I admit that I am rather rebellious in my free time and even if I have the costume, I am always the same, only when I wear the costume, people do not see it that way ... But if I show up with the combat boots, people think, aha, look, she likes Hitler, and she drinks, smokes, messes with people, and things like that. Us riflemen are portrayed that we only get drunk at the festivals, that we take drugs or God knows what, like Hitler and all sorts of things and that all girls get laid and things like that."

It doesn't matter to Dagmar: "Because it is the same to me when people think that I am so and so because I know what I am." She is not that well-behaved, but does believe in God, prays to God too. Around her neck, she has a pendant of the virgin mother. The scaring of people in church if they don't live in a certain way does not get to her anymore. She knows that she is not a "Hitler." She knows a few Hitlers, but just as casually as the people from the "Satan" circles in Eppan, Bozen, and other villages. She wears the combat boots with regular, black shoe strands, not as a symbol for anything. They are comfortable. They provide support when riding a motorcycle, "and in the city, the steel cap at the front is handy because it protects you if someone steps on your feet."

Her favorite place, where she feels at home, is the Giggelberg, on the Nassreither alp above Rabland: "You go up there by cable car, then you walk another hour. Then there are the flowers, which gives me a feeling like I'm fine again, and on the way up is the virgin mother, we put fresh flowers there when we go up, and then I sit and look around me, that gives me ... my sense of security back."

Between Angel and Skull

Summary description 2010; Interview on October 20, 2009, first on the way from work in Terlan to Lana, then in her mother's apartment (absent); Dagmar is 26 years old. She talks directly about what has happened in her life: break from her mother and her boyfriend at 15, moving out of home, pregnancy at 16, separation from the child's father, fight for the child awarded to him, debts, difficulties, broken educational paths, but still dreams and hopes for which she has her little niche *Heimat*.

> *"It seems to me when I feel really exhausted, just stressed or annoyed, what can I say, when there is just too much work or too many problems or anything else, I always have my place where I go, what I also need, right, because if I can't go there, if I need it, then you can never talk to me, honestly. [...] Yes, I go alone, alone or with my boy. Then we throw stones into the water at the lake ... Yes, I need that, I have to be honest because otherwise, I think ..."*

> *"Well, Italy, yes, yes ... you know, I always say that, according to my identity card, I am an Italian citizen, but in my eyes, I am and will remain a South Tyrolean. It will never change. But I also think whether South Tyrol belongs to Italy or Austria that in today's world, now, maybe that doesn't make the murderous difference anymore."*

> *"So for me, it is important that I see mountains in the morning when I get up. The rest doesn't matter, but mountains! It is brutally important to me. I don't know. It's like a protective thing or something. I have no idea. That you feel protected, maybe that's why I couldn't be anywhere else, I don't know. The mountains are high, are around you... something like that. Now is my favorite time in autumn anyway, when all the leaves and everything becomes colorful."*

> *"I have to be honest, in Lana, that's where I found it hardest to be ... When I look out of the window here, I only see the moon and the stars ... Yes, I don't feel really home-y, let's say ... Because ... yes, it is different than in the upper Vinschgau because, in the*

Vinschgau, it makes no difference whether you live in Rabland or Schlanders or Latsch or Goldrain. Here people are different; I just notice that people ... well, people have generally become more selfish in recent years ... But here, it seems that even if you have lived in Lana for 20 years, you would never be one from here. They let you feel that ..."

"I often fear that there may be a civil war here in 20, 30 years. Because I don't think there would be a Third World War. But if it continues like this, because people are always more dissatisfied, people always have to work and earn more ... [...]. Because it starts with social housing, the first-come from Pakistan and who knows... I am not a racist, so I mean, the people who come here, work here and pay and do everything, and when they get rent contributions, I don't care, that's okay, we get that too. But if they come here and don't even move their ass and drive the whole day in the fat Audi or BMW through the area where I think, yes, I'm crazy, I'm going to work as a rag and can't even afford a small car."

"Why are there so many illegal foreigners here? Someone has to let them in. Money has to flow to get there. Nobody can tell me otherwise. And with these poor people who work their whole lives to get on such a ship and then risk going on, they lock them up below. And the others they just let them in."

"Yes, I mean ... I have to say honestly, well, now I talk really stupidly, but ... that I ... have a foreigner ... yes foreign ... friends, that's only since I've been living here, [...] And you know why, because the people who live there and are from Lana and were born there and what do I know who got the blessing, I don't know, they treat me, yes, as a foreigner, you know.."

"So what I ... what I would be willing to do for my Heimat, so to speak. So everything without a child. If I didn't have the child, I would fight at the very front, but I can't do that. With a child, there is already a border... that is clear. So I would never put my child's life in danger or endanger myself through the child, actually. It would also be selfish. I mean, it would be to my child. If someone says, let's start a war, uprising or weapons and things, earlier I would have done it, but I just had to think of myself [...], but if I go to war today for my country, Tyrol, so to speak, then something is different because then it is already clear that you may not come back."

The fleetingness of her real homeland, the changing of locations, a quick dismantling of the tents without leaving many traces is also noticeable when searching for Dagmar in 2009. In Rabland, neither she nor her mother, who bears her single name, is in the phone book. Neighbors vaguely remember "the dark girl" with the black dog, yes she was called Dagmar, but the mother was not known, no idea where they were going, they had long been gone. The riflemen captain of Schlanders, where the daughter and mother were still members in their Rabland days, remembers both but is no longer in contact. He only heard that the mother had moved to Lana. But there is no entry in the phone book in Lana either. Nobody knows them from the riflemen of Lana. Only the registration office confirms that the mother and daughter are registered in Lana, but a written application is required to pass on the address.

Dagmar has no Facebook entry and, as will be shown later, no Internet connection. Finally, the only way to contact her is to write a letter asking her to call back if she is ready for an interview. There is no answer. Another internet search gives a single result on a woman with the same name as Dagmar's mother. As a Peter Maffay fan club member, she was at a meeting in Marling, which even showed a photo. The president of the fan club knows her and passes on her number. The mother already knows that Dagmar was happy about "the letter" and wanted to call back long ago, that she only came home late in the evening and would have no desire than to call anyone else. She doesn't dare pass on Dagmar's cell phone number because the daughter is sensitive. But she would remind her of the letter. After a few days, Dagmar answers the phone, in a good mood and ready for an interview. She had become a little older.

Once again, the planning and especially the choice of location for the interview are complex. She suggests a meeting point in Bozen, just outside the city center, where her father lives, where she sometimes stays. The day before the meeting, Dagmar canceled, something had happened, but we could meet in her mother's apartment the next day. I suggest picking her up after working in the nursing home in Terlan, where Dagmar now works. She agrees to this; from Terlan, we drive to Lana. Dagmar is bubbly while going. When we arrive at the mother's apartment, she is not there. Dagmar apparently had arranged it that way. She offers coffee and cookies.

Dagmar's story about the difficulties with her mother's former boyfriend and her early motherhood gives the conversation an unexpected twist. It is no longer about an attitude towards *Heimat*, but rather life events, which may have changed some attitudes as deliberately reflected by Dagmar. *Heimat* is a life story, one could say. For Dagmar, this means: At 15 a break with the mother and above all her boyfriend, moving out of the parent's house, juvenile court, at 16 pregnant, young mother and, after the break with her boyfriend, early, still a minor single parent of a boy, who then is judicially awarded to the (adult) father. In contrast to 1997/1998, Dagmar is also very open about family injuries; she only asks for protection in those aspects that could harm the child, i.e., as little detail as possible about the custody dispute and the child's father. Suffice to say that she could not understand why the child had been given to him and not her. She was neither a prostitute nor a drug addict nor an alcoholic ... After a long struggle, during which she hardly saw the son at all, the visiting hours for her were at least extended, the child psychologist had told the ex-boyfriend that the son also needed the mother.

She speaks lovingly of the child, usually lowers her voice, and chooses a warm tone. When describing beautiful moments during her retreats, she usually mentions her boy. She also reflects on upbringing questions, especially how to speak honestly to a child without shocking him. Yes, for the child, she dampens her other way of saying everything straight out and without sugar-coating, glossing it over a bit, beating a little more around the bush, but she is always honest. The "softening" comes close to the way of talking and not coming to the point that she chalks up to her father—not in a bad way—as "cowardice," as weakness, only with the clarification that she is definitely telling the truth, including with the child. She is at peace with her father. When she is in Bozen, she sometimes lives with him: "It is like this, I died three times for him and rose again from the dead. [...] Because I always felt that dad needed me more than my mom. Because dad drank, dad slept around, they said, and mom worked ... dad worked too, but I knew that mom was strong ... I always saw dad as weak."

She had already been open, direct, keen, aggressive in the interview in 1997, milder during the follow-up discussion in spring 1998 on the way from the school in St. Pauls to Rabland and then at home in the garden. But what was woven into the stories of girlfriends at the time is now straightforward: the conflict with the mother's boyfriend, the alcoholism of the (now sober) father; about the mother's boyfriend, if he was the same, she had said in 1997 that she had rejected the first, with the new one now, the "step-y," she had actually brought them together. A little later, according to Dagmar's description, he was the reason why she packed the bags and moved in with her boyfriend. The mother's boyfriend, in Dagmar's experience, had a hard hand with the stepdaughter that the mother "did not want to admit this"; from Latsch, she and her boyfriend moved on to Goldrain, a migratory bird, up the Vinschgau again, as she had previously walked down the valley as a child. It reminds you of the nomadic Korner almost inevitably, restless, without permanent abode, with changing locations, a homeless person who is always looking for a new "place."

Dagmar's lifeworld has several times come close to the existential stress that for Habermas goes beyond problematization, which is a complete breakdown: break with the mother, without support from the father, for whom she dies, again and again, pregnancy and motherhood at 16, separation, deprivation of custody, debts that are partly borne by well-intentioned guarantees, bursting education and training dreams. She wanted to be a mechanic but didn't get an apprenticeship, so she waited, helped out in kitchens, wanted to get back to training for social professions at the age of 18, but finally had to give up because she was mainly in the bar until late at night. "But I'm still in my head, nursing assistant or end-of-life companion ... yes that would be my ... top choice, let's say it that way."

She is 16, a young mother, works in a bar where an old woman who has cancer comes in, who is rarely visited by her children and who talks a lot: "Yes, and I liked talking to her, she was an intelligent woman, lots of wisdom and all that, all good advice too, and I actually liked to talk to her." The acquaintance became a kind of friendly terminal care, which brought Dagmar to her career wish: "Yes because I notice that many people are old or dying, alone, and I don't like that. Because I don't think anyone wants to be alone when they are dying or ... seriously ill or something. So I would like to work directly in a hospice [...] I don't know, it seems to me that you are doing something meaningful, which then gives me another fulfillment for me."

Then comes prison, a short but abrupt incision that once again throws all life dreams she had overboard. She experiences the humiliation of jail that she couldn't have imagined and experiences the protective hand of people who committed serious crimes. The arrest was her request that she be given back loaned money, which was interpreted as extortion, a charge that was finally dropped. Pre-trial detention in the Rovereto women's prison is the only experience that Dagmar does not talk about initially, which rather slips out. Still, she insists on explaining when I—already unsure about it and feeling a boundary—already want to change the subject. The fact that she names things strengthens her. She clarifies something that belongs to her, stands behind it. Can this be identity work by making exactly what others accuse you of, what could weaken you, the core of your own identity? According to Habermas, the problematizations and earthquakes in Dagmar's lifeworld seem to have made areas of her living conditions more conscious and communicatively accessible.

In 1997/1998, Dagmar still spoke primarily of political ideas about *Heimat*, of the fight that she wanted to wage for *Heimat* like Andreas Hofer, and even to the point of death. Speaking about *Heimat* in 2009 became a biographical reconstruction and an illustration of niches in the world primarily: "Well, I've become a lot more sensitive, oh my god yes. I do not know what it is. Yes, I do not know. Was it the pregnancy, was it just the time? I simply see many things differently." The political threat images, although changed, are still there and available; as soon as Dagmar talks about it, she talks herself into a rage again. But they require concrete inquiries. Otherwise, Dagmar would probably hardly have spoken of politics unless it was a detour through injustices in life.

Some striking key experiences offer themselves as moments of learning:

- the break with the mother, the attempt to build her own self-reliant existence
- the birth of her child
- the intensive encounter with an elderly woman in the neighborhood whom she accompanied until her death after her illness
- the experience of a social decline and deprivation of custody
- the experience in prison
- the opportunity to see her child again
- reconciliation with father and mother, the failure of whom she can talk about openly and free of accusations
- the experiences with people in the old people's home
- the foreign experience in Lana with friendships outside of her culture
- the friendship experiences with foreigners, the pen pal with a foreign prison inmate

Injury and processing characterize Dagmar's telling of her life. She portrays *Heimat* as a shelter. She is no longer protecting her *Heimat* with the riflemen, but *Heimat* is a "protection thing" for her. Her description of the mountains that surround and protect her arouse womb fantasies. She creates a closed circle with her arms when she speaks of the mountains. Her mother, she had told in 1997, almost died when she was born. This time she doesn't mention it anymore but speaks of extreme high-risk pregnancies that she would expect with more children. She criticizes the weakness of her father, to whom she feels more attracted, albeit lovingly and expressly "not pejoratively." In an admiring tone, she talks about the mother that she has "power," balls, testicles, and equips her phallic power that she may miss in her father, but who she can take care of. As early as 1997, she had spoken of the girl the father had simply forgotten when he should have picked her up from school and who washed the stairs for the father when he vomited on the stairs. The division of her parents into strong-mother/weak-father saves possible shifts, which she also talks about in other contexts: that the father is now sober, that he has an apartment in Bolzano, that she sometimes stays with him while the mother has developed mental problems, possibly also through Dagmar, but also because she has swallowed so much, swallowed in the sense of repressed.

Her image of women, which she claims is masculine, phallic. Just as she would have liked to have been a captain with the riflemen, would have waged war, just as she'd hit the mark in target shooting, just as she'd stood up against stronger men if they insulted her: the image of a woman who defends herself does not hope for male protection, however much she longs for it. For example, in the statement from 1997 about how lovingly the riflemen took care of her when she had problems—a surrogate family for her back then since she had a broken family. But she had to experience that there are always only a few people when there are real difficulties. For example, when she left the riflemen because of her boyfriend, she received no further help from them. Dagmar has learned to help herself and, as far as possible, to correct her mistakes herself. At workplaces where there were many "women," she noticed that they were just cackling around like in kindergarten, and even though she was much younger, she often had to tell them what to do. At the same time, there is a learning moment in recognizing a border where she could not progress on her own and was grateful to be caught by the mother.

Her political judgments are tempered. Without compromises, it would not be possible, going back to Austria or a free state would be lovely, but it doesn't seem so important to her anymore: "But I think to myself today, if that's not the case, then you have to make the most of it, then both have to step back and find a solution that works for both sides. I used to be very radical there, I have to say." Such a statement is remarkable because it refers to *Heimat* in a political context, at the level of "system." In discussion with Habermas, the problematic experiences in the lifeworld have not only made it more conscious and configurable but also influenced the perception of the system. It added mitigating, balancing, enabling ambivalences to the system, allowing moments of communicative action to appear at least theoretically: "… finding a solution that works well for both sides".

The political phantasms have not disappeared: the threat images of foreigners who drive up in big cars, do drugs, and do not work, the Chinese who buy all the shops in Meran, and finally the fear of civil war because "the people stand up" are dark shadows over these lighter thoughts. An obvious explanation is that disposing of the suffered psychological and physical violence into phantasms and projections. On the other hand, Dagmar is reflexively aware of all her injuries, pronounces them, sometimes even accepts them, and has done both mourning and reconciliation work—as far as mother and father are concerned. That may have made the reflection and exploration of her lifeworld possible, moreover forced by the sometimes existentially real-life twists and turns in her life. Not everything may be reconciled, not everything cured, which would ultimately be a naive idea. Dagmar has a second job to pay off debts. She can only afford a little, is a motorcycle and car fan, is interested in tuning engines and bodies, but has no car and could not even get a driver's license due to lack of money, goes to work bus cumbersomely. Because there is no other way, she lives with her mother at an age when others are moving out: "Yes, there are a lot of things that I knew if I go home now … then I lose 40 percent of myself. But it just had to be."

With her debts and after prison, it was the only way to get back on her feet. As powerful and energetic as she is, as joyful as she can talk about fishing, throwing stones at a stream, as trustingly she prays to Our Lady and God, socially she is an underdog. She is, ultimately, that people of whom she believes that one day it will rise out of poverty and rage against the rich and start a civil war, but which she would not join—for the sake of her son. If she didn't have the boy, she would be right at the front again.

In Lana, she is a stranger, the riflemen don't know anything about her, and maybe they wouldn't want to know anything anymore. She perceives the locals as snooty and marginalizing. If *Heimat* has something to do with social integration in the place where you live, then Dagmar is homeless. Her only social contacts and friendships are with foreigners, the Muslim Rita from Kosovo, Diana from Slovakia. In prison, she experienced the protection of an Italian-speaking migrant. When she was released, and the other woman had to stay, she cried; since then, she wrote her letters in Italian. When she talks about Rita and that this relationship is a give and take, not just one-sided exploitation of friendship, her voice almost breaks. But she sticks to the stereotype "foreigners," only admits that there are good and evil, just like the local ones. In *real life*, in the *Heimat* as lifeworld, she experiences foreigners as reliable friends, is grateful to them; in the *Heimat* as a system, however, the foreigners still disturb, they appear as a threatening danger.

When asked about the service to *Heimat*, which was central to her time with the riflemen, she lists almost only lifeworld-related topics (self-help groups, summer kindergartens, care for the mentally ill, AIDS help). After the interview, she also says what needs to be changed: take the bombs from all countries, programs for young people, education about the dangers in car traffic. A lot has to do with caring for her child. But the effort for education and learning also plays a role. Her educational paths have all failed for the time being, but even in failure, she has taken a lot with her and developed some ambitions. She often goes to libraries in Meran to borrow a book. Technical terms from social work such as "family background" or "integration" come a little unsure across her lips. She gets stuck—otherwise speaking fluently—both times in the middle of the word, but she knows the terms, pronounces them.

There is a struggle in her: Dagmar projects images of threats that she rejects for the sake of her son, she uses the example of a friend from Kosovo to describe the cruelty of the civil war, but if she had not had the child, she would march ahead into the war. Dagmar was forced by the problematizations of her lifeworld to deal with it and to project her difficulties less than before onto enemy images. As a result, she found expanded and relaxed identity possibilities, the fruits of which were primarily denied to her in real life. The custody of her child is her greatest hope. She believes that she will have paid off her debts soon, that her dream of becoming a palliative nurse may come true after all. None of this is there yet, she is holding onto the dreams, but her fear and anger do not need to be disposed of until the dreams are fulfilled. So, in a precarious present, she clings to her images of the enemy and intensifies political criticism because she sees in the Italian government and cowardly South Tyrolean politics all the evils of this world gathered and therein can dispose of her own. Even the realistic view of her life that she has fought for or that has been wrested from her life sometimes blurs. This is how she describes the binge drinking of the boys, their neglect by the parents who drink themselves in a way as if this were a problem only for this new generation and above all a problem that others have. But it is pretty much a description of her misery in childhood and adolescence. An increase in awareness can also be recognized there. In 1997, she resisted the riflemen being labeled a drinking club. In 2009 she confessed—albeit with a slight smile and a trivializing choice of words—that they would really "booze." Boozing is different from drinking, as perceived by "today's" youth.

She only kept one memory of her time with the riflemen: "my mini costume when I was tiny." On the Sunday of Sacred Heart, she sits at home and looks out the window, but she cannot see the mountain fires from her apartment. So she lights her own Heart of Jesus fire with candles and explains to her son what the mountain fires are all about. Yes, she laughs, that might be ridiculous, but it gives her something, as she keeps lighting candles in chapels and thanking god every day in the morning, "that I am healthy, that my boy is healthy, that I can go to work, that I have a roof over my head and so on, and I do the same at night." That is what she says, although she does not like the church with its moral constraints and hypocrisy. Then a pattern returns that had already shimmered through in 1997, a rebellious trait, a commitment to individuality and honesty, also in moral questions, combined with disgust for the assumed deception of others. "The fact is that mass today is only a fashion show, and everyone gossips about the others, in front of you, behind you. Some sex talk, the others talk about the neighbor, how he screws everyone, the others talk about that … so I think that shouldn't be in the church."

From "back then," she still has a soft spot for skulls. Her cell phone, which she uses to send photos of herself via Bluetooth, is called Black Angel. The black angel, she laughs, is herself but surrounds herself with guardian angels in the apartment. Death is a strong motive in Dagmar's statements. She would be ready to die for the *Heimat*. She could pass away like her mother when she, Dagmar, was born at future births. She venerates Andreas Hofer, whose myth finds its most remarkable unfolding in victim death. At a symbolic level, there is also a longing for death in Dagmar's relation to Our Lady and the guardian angel. With the skull cult, she also banishes everything that scares her. Among Dagmar's possible learning moments are two life-related debates with death: the many conversations with the dying older woman while still working in a bar in Goldrain, now the desire to work in a palliative care unit.

Her son's name is Andre, like Andreas Hofer, but in French. Dagmar has researched her family history on the paternal side. Lafler came from Lafoglèr and even earlier from Lafollie or Laffollier, the family descended from a French soldier who stayed in the country after the fighting against Andreas Hofer. Dagmar has learned to see some things with a grain of salt.

Johnny Ebner

Heimat ... that is Texas

Johann "Johnny" Ebner, Lüsen, riflemen company Lüsen, 22 years old; Interview on August 31, 1997, in Goldrain, in-depth discussion on December 13, 1997, in Lüsen and mostly in his room. Two meetings followed on December 20, 1997, in the riflemen club and on May 17 at the soccer field in Lüsen.

"Heimat is a lot. It's a lot. I'd say ... Lüsen, first of all. First of all, Heimat is my village Lüsen for me, at home, South Tyrol certainly too. But not the way they often make it sound, just ... let's say I'm not that fanatic."

"I have a very different attitude. I am ... I am also a football fan, a massive one, and nationally I am Milan fan ... and I have no problems with that."

"There are often fanatical ones among them, but not here in Lüsen. We have a brutal cohesion, so nobody talks stupid because I'm a Milanista and the other is a Bayern fan or something. Sport and politics are very different."

"For me, the costume is ... Zeilinger ... leather pants, a Tyrolean! [...] Male? It is a bearish ... expression of ... Tyrolean. It's old. It's tradition, something amazing. The fire department also has an outfit, right? I don't know what they think. I think it is ... it's the tradition with us, anyway, the only one, right? Yes, the music is still there too."

"It would be good to protect something! We are certainly there to protect something, but nowadays ... I do not talk so much about national defense anymore, but see more that there is a little there ... that everything works a little in the country. Rather see

that ... we have a celebration here and there or something, I'm more for it. Defend? Well, if we need it, I'd do it again. [...] Yes, you couldn't act otherwise if it is like in Yugoslavia. Then you would also stand by your country. It's just the way it is."

"The rifles?[9] It would be nice, but ... better than walking around with a wooden rifle, I'd rather walk around without a rifle. [...] Well, I don't need a gun. We can also have a beer afterward."

"Simply extreme." "Texas." A *Heimat* lies hidden between these sayings. A bakery as a local political achievement. A village at the end of the world. Driving down to the disco, park the car so that you don't have to tap into the alcohol check when moving home in the morning. Cycling on the toughest trails. Playing soccer, being a soccer fan. Down to Italy to the big stadiums. AC Milan, "this is the greatest." Johnny is having a beer with Helmuth and uses a jocular tone. He becomes serious when it comes to his demarcation from fanaticism and violence, describes his mother as a role model: she was a heroine because she was "really on the ball," "genuine." One room: "Freezing cold in winter. You don't believe how cold it is there. You have to be hard. Everything has to be hard with me. If it's not hard, I don't make it. I'm a baker. They're the toughest; you don't believe that."

Lüsen, upper Brixen, or not? It is a place in between. The Lüsen valley flows westward into the Eisack valley, near Brixen. The Lüsen stream, called Lasanke, rolls eastward into the Puster Valley. So, where does Lüsen belong? The people from Lüsen say a simple "we are Lüsner." This gets them around a location determination. Lüsen is something remote. The Lüsner say it is something of its own. "Lüsen is brutal," says Johnny Ebner. "Extreme what we drink there." The village has the highest beer consumption per inhabitant, he believes. It is said to have the highest alcoholic rate in Italy.

The Lüsen village book (Delmonego 1988) is the chronicle of a race to catch up. The village was always behind. Because it is secluded, every failed harvest meant hunger. No stone was left unturned when the torrent Lasanke broke into the town. The construction of an access road in the First World War brought hope. A village fire in 1921 destroyed it again. Until the end of the Second World War, nothing happened in Lüsen. The youth migrated, looked for places in the

---

[9] Riflemen in South Tyrol were forbidden by the Italian authorities to carry historical weapons. While their comrades in Austria and Bavaria fired salvos at festivals, this was denied to the South Tyrolean companies, which caused constant debate, as it did at the time of the first interview series in 1997/1998. In 2000, the South Tyrolean riflemen were finally allowed to carry the historical guns, but they had to be dummies that allowed a loud bang but no shooting.

city down south, or drowned themselves in alcohol. Johnny attended vocational
school for trade, then worked in a C&C supermarket in Brixen. He couldn't
stand it there. Better to be a sidekick on construction sites, bricklayer: "I need
something hard."

Lüsen got sports facilities later than any other village in the up-and-coming
South Tyrol of the 1970s. Johnny plays football for a leisure team that he founded
himself, rides the toughest routes, and has an exercise bike at home. He has to
make sure that he is always in a good mood. He is in a good mood 320 days a
year, he estimates. And the remaining 45 days? "I'm delivering a brutal one."

Most of the time, Johnny hangs out with Helmuth. He joined the riflemen with
him when there was talk about the riflemen in the village bar. Someone called
over to them. Do you have anything against the riflemen? "No, we have nothing
against the riflemen." "Then you have to join them." Then they joined. Johnny
has made only one condition: not a word about him sticking to an Italian team.
"Milan is a matter of honor!" Helmuth is a Bayern Munich fan. But Helmut is
also the only one who can call Johnny "Hons," the South Tyrolean pronunciation
for Hans. Otherwise, he can't take it. Johann is on the ID card, but he is Johnny,
except for Helmuth.

The construction of the bakery where Johnny works is recorded in the village
book, so so important was this. Until December 1990, the people of Lüsen had
to have their bread delivered from Brixen. Now they have their own daily bread.
Johnny first helped count bread. Then the baker asked if he wanted to be a baker.
Now he is mixing the dough. "This is the hardest job when baking. The bakers
are the toughest, brutal, even with the girls, they always have several…, even if
it's not true", he grins. A working day from eleven in the evening to eight or
nine in the morning, then sleeping until two, three, then getting up, cycling up
to Afers or down to Brixen, and then back. When his colleagues go to the disco
and have to go to work, he gets annoyed sometimes, but then he thinks, "I'll just
make a good loaf for the people of Lüsen."

Politics, no … Italians are also not an issue … there are different types. That's
just history … "There were things that happened that were brutal." He says it
hesitantly as if he should consider whether to give it away: One of Johnny's
uncles was one of the "Pfunderer Buam," "the first to be locked up by the Italians." The "Pfunderer Buam" were young fellows from a small mountain village
in the Puster Valley. Their story is a sore point in the South Tyrolean memory,
which was one of the triggers for the violent struggle for autonomy of the 1960s:
they had gambled in the local restaurant, financiers were there too, at first there
was joking and fooling around, financiers and fellows alike, then there was suddenly quarrel over the curfew. They quarreled and left the restaurant. There were

fights, and the financiers ran away, one, Raimondo Falqui, probably plunged into a ravine alone or in a hand-to-hand battle. It was so dark that his colleague didn't even notice. It wasn't until the next day that Falqui was searched that he was found dead in the stream. Everyone involved in the dispute was arrested, towed in heavy chains, accused of willful murder, and sentenced to extreme lengths. "But let's not talk about it," says Johnny, "we don't talk about it at home." A cousin married the "Ciano." This Italian man is a veteran of Gladio, the notorious secret service suspected of connections to Italian terrorist groups and sabotage acts against the South Tyroleans. But no one in Lüsen believes the Ciano can do anything wrong, "he is a Mushroom picker, a brutally passionate one."

The former farm where Johnny lives is called "Mühlhäusl," the small house on the mill, and it is tiny. A stone that the Lasanke had washed up was as big as the house, says Johnny. There is no livestock, a little forest, a small potato field in front of the house, and a vegetable garden with chives, which his mother planted and has survived her death. The farm was Johnny's *Heimat*, he thinks, as long as the mother lived: "She was really on the ball, just genuine ... she stood by what she had and always helped you, she cooked well and ... she was a superwoman." Now only the room is his *Heimat*, so we should definitely see it the next time we visit: "You won't believe that when you see it."

Johnny has hidden the key to the room. Only the sister is allowed in, nobody else. He gets along well with the brother. He pays the electricity for the whole house, which he likes to do, "as a baker, you earn very good rubles." He has a Golf GTI with a CD player. He records CDs for the boys in the village on the stereo in the room, labeled "John Ebna-Mix, perfect." He had a girlfriend, but she left him. "Got over it, stayed the same." He folds his arms as he says that.

The room is cold. You don't think how cold it is, he had said. "So what do you say now? Cold, right, hard." The room is a spectacle, a stage covered with soccer flags, everything black and red, AC Milan. Johnny fetches bows and pennants from the wall, passes them to the touch, pensively strokes them, and gives the place and date of the creation of such valuable pieces. "These are sanctuaries for me, you have to have them around first, then you know what's going on, bought originally, in the stadium in Milan downstairs, you put them on and go into the stadium, perfect!" In the car, he has a Milan flag as well, "just in case." When he comes home, he sits on his bed, turns on the stereo, "then it starts. Texas".

In early summer, we visit Johnny a third time. We'll find him on the soccer field, he says, training. His leisure team will have the first tournament in a few weeks. But only a few who had promised to come are there. "The boys don't have it. They train too little," says Johnny. He is disappointed, "but it will work," he does not give up, "definitely not. You have to be in a good mood in a village

like this". At the Italian championship, "Milan" was only tenth. They tease him
about it in the village, but he stands by it. He says it in the same tone as riflemen
talk about *Heimat*, about Tyrol. "Milan" is serious.

There is a drum in the room. He is the only drummer among the riflemen,
a second would be needed, but no one can be bothered. He does not let his
courage be taken away, "I just drum on my own, the boys cannot be relied on."
A Yoseikan Budo coat of arms between the Milan flags. "You have to be tough."
An open closet, the laundry folded, even the socks neatly arranged. He does it
all himself, including the laundry, also ironing, he also does the cooking, "once
you've lost your mother, you know what's going on." That is hardness too. He
got the vacuum cleaner from his brother, "it works." He put together and hung
up a puzzle with 1500 individual pieces in three days. There was one more, but
he stuck a few pieces wrong because they also fit in the wrong place, which
annoyed him terribly. "Well, you can see it when the light falls on it." A corner
is full of soft toys, which he collects and likes. He also has a giant beast from the
football club "1860 Munich", he put on a Milan hat, he christened it Giacomo,
"that is not German, and yet a little bit Walsch fits well." He has three earrings
and wears the mother's wedding ring, shows it in awe, nods: "You don't say
anything anymore. That's Texas".

Healing Wounds also Means no Longer Having to be Hard

Summary description 2010; Interview on October 3, 2009, in Lüsen in Johnny's
apartment, where he lives with his girlfriend Priska; he is 34 years old. The
parents' house burned down, Johnny built it up with his brother and set up an
attic apartment, he gave up the baking profession because he was "too hard," he
left the riflemen. He works for the Catholic Association of Working People.

> *"Well ... simply ... somehow with ... since the rifles came, I didn't ... it wasn't my thing
> anymore. Well, I don't need a rifle to be a rifleman ... [...]. It didn't really do it for me.
> From my point of view, it didn't fit so well anymore. You know, then they would give
> you trouble, why you don't have a rifle and all that. And I could no longer identify with
> it."*

> *"Also ... you know ... just that the riflemen, like they owned the whole show this year,
> with the crown of thorns, that does not excite me. Yes, Italy or not, I already know the
> history and everything, I know how it went and what was there ... But it seems to me
> that you should let it go at some point, which I ask from both sides, you know, because
> once one of them stops, then the others start something again."*

"I get along well with everyone. Whether there is an Italian or a German, I really don't care. I can get along with everyone, also with a Black or Chinese man, I really don't care ... As soon as it comes to politics or is about ethnicity or anything like that, then I tune out ... then I just switch off. When they talk about the Walsch here and the Walsch there, I just never listen."

"It just doesn't matter where South Tyrol belongs, it doesn't matter to me, I probably have Italy in my passport, but if Uruguay were there too, I would be the same."

"Yes, when I see a Tyrolean flag, I think of Heimat, yes red and white, then I think, that is neat. Or the Peitler Mountain, which is also nice, but there are other mountains too ... Otherwise a Heimat symbol? I don't think of Andreas Hofer and these people. Milan! Well, this is not Heimat, that is ... Red-black is beautiful, but ... No, when I see a Tyrolean flag, it is something nice, I like that. This is Heimat."

"What is not part of Heimat? Radical stuff, everything that has to do with things that are exaggerated, that is not part of it, politicians ... Political things, everything where it is exaggerated does not belong to the Heimat, you only do damage to the Heimat, be it political, be it economical, wherever things get exaggerated ... More down to earth, that's ... my way."

"Getting up, working, that is service to Heimat, I pay my taxes. That's already service to Heimat. Service to Heimat otherwise ... uh ... Service to Heimat? Service to Heimat? Next, next question."

"What annoys me the most, what I saw, that the Italians put up monuments of those killed in the war that is not true at all ... As I read this, you know, that these deaths didn't even fall there, that they brought them there. I think they are just doing it maliciously. It can't be otherwise. Why do you have to bring something here? That annoys me. But ... then I think to myself, let them do it, they will be happy."

"If the day starts well, that's important to me, neat, after a night with each other ... watching TV, those are nice feelings, going to bed together ... I like her, what more can I say. It fits. Next question."

"The main thing is that I'm healthy. It's just something important. [...] You know ... we are healthy and have food."

Johnny no longer says "Texas." That is the first thing you notice. He combines it with the fact that he is no longer "into alcohol" as he was with the riflemen. He still says "neat" or "it fits," sometimes "perfect." He also no longer says that something has to be "hard," that he has to be "hard."

Hard for Johnny 1997/1998 may have meant: not allowing yourself to be weak, not getting carried away by the injuries, not even being afraid of where he was going, and being ultimately alone. He was with the riflemen but didn't fit in unless it was about partying; he tried to keep a soccer team together "on a recreational basis" and had to watch them dissolve. He went on extreme tours on the bike to feel that he was tough, that he was "making" it. He delivered a "brutal one" to get over the fact that he wasn't in a good mood on some days. The survival law for Johnny was to be in a good mood. In addition, at 22, he still had cuddly toys in his room and Milan as a substitute home.

His siblings only appear in grazing lights. At that time, the sister to whom he entrusted the key to his room and the brother who had given him a household appliance; this time only the brother with whom he built the house. The father, too, does not seem so important: when Johnny was a baker, he could not sleep in the morning because the father "opened the wood" below; now Johnny is chopping the wood. The size of the farm was not sufficient for being a farmer, so keeping animals (formerly a few cows, then only rabbits and goats) was more of a quirky hobby for the father: "But that didn't do it either, they eat everything, these whore creatures, well, well ... [...] It was just a burden, it seemed to me, he is not able to keep the cattle in the barn, but you can't let goats run around, they eat everything, eat bark from the trees, the wild ones... And the same thing happened to the rabbits, there were rabbits everywhere, and then the fox got them all." As an employee in torrent management, the father had a job that allowed a lot of free time in some seasons. This may have given Johnny the feeling that the father had passed the time with senseless animal husbandry while the mother was "really on the ball," as he said in the first interview in 1997 when asked about a role model. Her early death made him silent when he visited his room: "Then you know what's going on."

Even now, what he says about the mother is kept short and essential. He points to a picture of her on the wall, nods, says that it is all right. He is still visiting her at the cemetery. At 20, when she died, Johnny was no longer a child. The wound that the mother's death may have torn open may have been something that was already there or went deeper. His fixed idea to be hard, trampling down weakness, doubt, fear, or being alone indicates this. Having to be hard is close to inflicting pain, maybe to feel it, to make sure it exists, but it goes beyond it: Because it's hard, that I can endure the pain and the stress, whether cycling or in the bakery, I not only make sure of my existence but also that I can do it.

Johnny came to the Catholic Association of Working People not very differently from how he got to the riflemen: someone had spoken to him, he had participated, and then he had been elected to the KVW committee. But it is also the change from an association (the riflemen) in which *Heimat* is strongly present as a thought "system" to another association concerned with the life-world's design, to make it more beautiful for others who are otherwise alone or struggle.

The visits to the retirement home are experiences in which Johnny encounters the mostly repressed fear of growing old, frailty, and death. He calls the "Jesu-Heim" in Girlan "Herz-Jesu-Heim" (Heart of Jesus Home), wrestling the heroic political myth in a Freudian slip of the tongue, a life-like confrontation with human ailments. Johnny's commitment to wanting things to be okay, to live with a few pleasant aspects, healthy and with his girlfriend, without great demands, but with his own joys and social responsibility, is a commitment to *Heimat* as a lifeworld. Everything else, for example, South Tyrol or the flag, still has a meaning, but the questions of where *Heimat* begins and where it ends overwhelm him, making him perplexed. He has no connection to it. *Heimat* is here. He knocks on the table: the table, the apartment, the girlfriend, what they might both make of it.

In his search for his life opportunities, Johnny made a clear decision: he left the riflemen without a fight, but feeling that it just wasn't for him anymore. He returned the costume without a problem. The leather pants had belonged to him anyway—and that is the most important thing, he can laugh again. He logically spends less time with Helmuth because the friend is on the go with the riflemen, while he is more likely to play sports or be with his girlfriend. But "when Helmuth drives past outside, he often makes a stop. So contact ... no longer that much contact, but ... it fits."

The reason for Johnny's exit was the rifles, about which he had said in 1997 as a first reflex that they were "quite beautiful," but then added immediately that he did not need them. Nevertheless, it took a long time: Johnny only left the riflemen around 2006. The rifles had been given to them six years before, in 2000. When asked what the most important moments had been for him in the past twelve years, he came back to talking about the fire with which he had spontaneously started the conversation: "We were quite surprised ... I'm am on my bike and see that ... that the house is on fire, I thought, what's going on now, the mill house. We looked at it. We're both young, we never think of anything like that."

In light of the importance given to the house in 1997, the fire was like a collapse in the lifeworld, speaking with Habermas. With the destruction of the actual living space, all memories kept in the attic get lost: "Old stuff from the father, from the mother," then school notebooks, books, "a lot, yes, from the mother, mother-father, postcards, everything has gone." Johnny also dispels pain with jokes and quick exaggerations: You no longer have to go to the recycling center because there was nothing left to recycle from his memory. Then he adds: "Too bad," also because he can see from his girlfriend's photo albums that at home he had fewer memories anyway: "It's a bit of a shame that everything is burned."

The collapse of the lifeworld was also an opportunity for reconstruction: "Then you were forced. [...] Now we have built that up, and now there is, of course, much more behind it because I know that I did it somehow, as I said, together with my brother."

It is difficult for him to reconstruct the exact date of the fire. He initially connected it with the fact that the friend had studied to pass the Matura exam because they knew each other before the fire. Then he feels uncertain about this because that would have been in May or June 2000. But at the time, he came from the new job, not from the bakery in Lüsen, but the Zingerle in Schabs. And he only started at Zingerle in autumn 2000. He checks it and corrects himself. Yes, it was in June 2000; only then did he begin at Zingerle. But he was no longer a baker when the fire burned. He had to have his nose operated on because the flour dust "stuffed up everything." He mentions that only when trying to reconstruct the date of the fire. So the fire, as an event that shook his house, the fire also covered a personal shock: not being able to cope with a job with which he had associated toughness and his social role in Lüsen ("I'm just making good bread for the people of Lüsen"). He says again: "Yes, a baker is just a neat profession, that's neat. I like that. But it didn't fit anymore."

Forced by his health problems, which also led to chronic fatigue, he gave up the job he had chosen and returned to what he was before: "Store worker... you know, I've done a thing for a long time. Before being a baker, I was a storekeeper in Vahrn for three years. Then I just did baking, eight, nine years in Lüsen and ... It was just too strict, it is strict, it was just too strict, I started looking like a ... like the white wall there." With the bakery profession, forced by problematization in his life, he also gives up the hard work at all costs. He is already looking for an even easier job because, as a store worker, he often has to carry heavy things. Johnny allows himself the weakness of not being able to be a baker anymore, not being as brutally hard as he thought was indispensable in 1997. The fire in the house forces him, together with his brother, to take charge of his life and rebuild something that was destroyed, where the mother's death had probably broken something. The retreat to his room, decorated with Milan flags, had something of strangeness in his own house. By rebuilding the house, he probably also heals this strangeness, makes the house his own, together with his brother. He sets up his apartment on the top floor, where memories of the past have been burned. Viewed as a symbolic act, in the sense of a scenic understanding, the build-up after the fire in Johnny's rupture of life heals his fear of not being able to make it. He gives the Milan flags to young people in the village. A small closet with a pennant and a few coats of arms are enough for him.

With building a house, he managed to achieve something: he met his girlfriend, changed his job, created a new home, everything approximately in the year 2000: "Yes, a lot happened within a year," he says, "that was an important year, a lot happened there." From his temporal reconstructions, it follows that he started running around that time. First, he made "mountain runs, up there to the cross, then soon the marathon started." Unlike when he was cycling at the time, he doesn't say that running is "hard." He talks about how he immersed himself in books, how he started training and realized that he was going to make it as he did before, sometimes having a more leisurely breakfast before a race when "she" is there, the "dumpling" or the "scallywag" as he calls her. And he says that when he runs, he gets into that state of consciousness in which he no longer thinks and frees himself from everything, runs free. This is not "hard" or "Texas," but "neat" or "beautiful."

Johnny does not deny that something can irritate him about Italians at the post office who do not speak German or in the military if they let the tricolor light up. He also addressed the family-troubling chapter with the Pfunderer Fellows, which was taboo in 1997, one of whom was his uncle. But now Johnny can classify the dark side of family history, ironize it and leave it behind. Milan is no longer a substitute home. It is joy, fun, a hobby—but he no longer compares the Tyrolean flag and Milan pennants. If so, then the Tyrolean flag is still a symbol of *Heimat*—and the mountains, especially the Peitler Kofel above Lüsen. But if the patriots exaggerate again, Johnny is ready to tell them what's going on. He says he has learned the essentials from history, namely that the "exaggerations" were usually full of suffering for those affected. In the Tyrolean commemoration year 2009, Johnny did not take part in the patriotic parade but instead visited the state exhibition "Freiheit::Labyrinth" in the Franzensfeste fortress, which was looking for an accessible approach to the issue of freedom. The drawings on display, sent from children to the imprisoned *Pfunderer Buam*, brought him closer to the painful subject in a conciliatory way.

Johnny speaks casually throughout the interview, jokes, and confesses embarrassment. He seems only disoriented to the point of not understanding the questions when it came to the political-geographical demarcation from *Heimat*. While he quickly names the "millhouse" when asked about the "closest *Heimat*," Johnny becomes perplexed when asked about the "widest *Heimat*"—or rather, thinks he is at a loss:

Johnny: I feel good in a soccer stadium ... or what do you mean?

The greatest possible, wherever you say, I also feel at home.

Johnny: It's not that important to me...

Priska: If you are somewhere and go back to where you are from ... it is probably very easy.

Johnny: And what would that be for you?

Priska: Well, when I drive over the Brenner, I'm at home because I'm here in South Tyrol.

Johnny: Are you at home then?

Priska: Yes, also from the south ... but I was still in Italy then, I'm always at home there... But from outside, when I'm over the Brenner or the Reschen Border or the border in ... Innichen, or San Candido.

Johnny: Well, I don't think so, that's not important to me ... no, I'm sorry, I can't answer, well. *Heimat*? Nah.

Priska: Do you feel at home when you're up in Germany, Hamburg, and South Tyrol?

Johnny: Yeah, why not if I like it ... yeah, it's not a lie if it's nice ... The main thing is that it is neat. If the right people are there, then it is neat...

Priska: Yes, it is neat somewhere else, but then you will be glad when you come back when you ... come over the Brenner.

Johnny: I can't say.

Do you have something like an inner Heimat?

Johnny: I think ... let's say ... I find rest when I run, I see a rest, I can switch off if I know I'm going to run for an hour, then I'll add another hour then you get into something like that, then you switch off, you know, as if the brain switches off, then you only do more running, you do nothing, you only run, just that, you don't even think when you sleep, but when you sleep you dream, when you run, when you are in this thing, then ... Then I think to myself, I totally switched off, not thinking, that is neat.

What definitely belongs to your Heimat...?

Johnny: Can you repeat that?

What definitely belongs to your Heimat?

Hm.... *Heimat*? ... Now I'm under pressure, now I can't think of anything.

Priska: But it's very easy.

Johnny: Yeah, I am overthinking...

The question can also mean nothing to you.

Johnny: Well, maybe because you don't pay enough attention to it, that's right...

You don't have to say anything.

Johnny: Belonging to my *Heimat* ... in any case ... yes family, then ... my girlfriend's family, my girlfriend... Lüsen anyway, the apartment ... the work, work at home too.

With the transition to his own lifeworld *Heimat*, Johnny gets back on track, just as he answers the question about the "inner *Heimat*" without hesitation and individually about his trance while running. His blockade stems from the fact that he simply cannot empathize with political-territorial constructions of *Heimat* and sees them as artificial, imposed impositions, which they ultimately are. Establishing yourself in a small world that largely works without projections of political enemies or fear images, that is satisfied with a good breakfast for two, with running, with the search for a more comfortable job, with being together, is not necessarily a retreat. Johnny talks about how he is affected by television documentaries about world hunger and how much bread is thrown away in Austria. "That pains you ... but of course, we are too weak, but our politicians could go there and say, we'll come up with something." For example, he would see that goods in the supermarket are not thrown away just because they have passed the expiry date by one day, or "just because of a rotten apple that all others in the box that are still good are thrown away. You could give them to the homeless from Bruneck or Brixen. There would be a few who would be happy with it." There is probably the profiteering behind it, and that leaves him doubting the seriousness of politics, which leaves so many problems unresolved: "That gets me to one point.... They all talk smartly, but once you get the first "paganote",[10] they give you peace and quiet."

No longer having to be hard means living with inadequacies. To be able to say: "There is nothing I can do about it. It is terrible." Not looking away, but coping with the fact that some problems are too big and that you have to be happy to have the millhouse, AC Milan, the old people at KVW, the "dumpling," and health.

---

[10] Paycheck.

Knocking on the Door of Closed Traditions *The Overlay of Autochthonous Ethnicization and Migration*

The previous youth portraits represent examples of nationally and ethnically founded identity formation. The narratives of the young people, as different as they are, represent processes of identity formation between adaptation and emancipation within the own culture of the majority. Such majority cultures can also develop minorities if they constitute majorities over other minorities. And unfortunately, minorities are not so easily able to learn from their own experiences of suppression when they face other minorities or minorities within minorities as majorities.

What applies to South Tyrol, which Italy annexed in 1918/19, is what the feminist theorist Judith Butler in conversation with the postcolonial theorist Gayatri Chakravorty Spivak describes as the perfidy of the nation-state. The nation-state forces the suppressed and/or marginalized minorities into the ethnicization rejected simultaneously through the nationalization of its policies (cf. Butler and Spivak 2007, pp. 30 ff.). The national assimilation pressure with bans on language and culture forces minorities to narrow down their political articulation to national defensive struggles, to constitute themselves as a homogeneous ethnic subject and thereby suppress inner diversity, social, gender, sexual, and/or other differences. The cultural unity thus produced gains the strength of a protective armor over other groups.

The attribution of being different due to a few features that are not intended for it, such as language, origin, or religion, leads on the one hand to a discriminatory external ethnicization, which Mecheril (2002, p. 107) tries to grasp with the concept of *Migrationsandere (migration-others)*. The term should express that immigrants are not different per se but made discursively into others through the perception of migration. At the same time, this external ethnicization can interact with practices of self-ethnicization, with which minorities form communities and—both personal and political–attempts to become subject matter. In this

© The Author(s) 2023
H. K. Peterlini, *Learning Diversity*,
https://doi.org/10.1007/978-3-658-40548-9_4

way, ethnicization has an ambivalence between discrimination and empowerment, particularly in tension between foreign and self-ethnicization.

Suppose ethnicity is understood on the one hand as a deviation from a discursively claimed form (foreign vs. domestic, black vs. white, Turkish vs. German) and devalued accordingly. In that case, the corresponding exclusion goes hand in hand with the generation of belonging. Thus ethnicization offered the German-speaking South Tyroleans, who had been severely discriminated against as a linguistic minority in Italy for decades, the opportunity to form themselves as an ethnic minority. In this way, they became a political subject (at the price of external demarcation and internal homogenization). Similarly, the discriminatory exclusion also stimulates migration groups and diaspora communities to define themselves as "Turks" in Germany or "Bosnians" in Austria and offer their compatriots appropriate affiliation, even if only to a disadvantaged and – negatively connoted—group by the dominant majority. Ethnization ultimately expresses itself as a counteraction to tendencies towards assimilation, i.e., becoming as similar as possible to the dominant majority in the target country (cf. Peterlini H.K. 2017b).

For migrants who come to South Tyrol, this problem arises even more complex. In supposedly homogeneous national societies, "due to different characteristics—external or linguistic – a different ethnic background is attributed to them" (Huxel 2014, p. 71). In a majority-minority area like South Tyrol, the normality umbrella is now without national uniqueness, with which alignment or delimitation can/must be carried out. People who immigrate to South Tyrol sometimes only gradually realize that they have arrived in an ethnically divided reality—on the one hand, Italy with its state structures, on the other hand, the Autonomous Province of Bolzano South Tyrol with a division of almost all public areas by language groups, namely German, Italian and in the two Dolomite valleys also Ladin. Kindergartens and schools are separated by language (e.g., the German school with Italian as a second language and vice versa), public offices are allocated according to an ethnic distribution key, the funds for social housing are also balanced according to language groups (Peterlini H.K. 2017b, p. 156f). The formation of identities among young people is thus ethnically established. Young people find it difficult to evade the pull of ethnic separation in forming circles of friends and selecting assignments (cf. Chisholm and Peterlini 2012).

For new arrivals, this means not only having to move between pressure to adaption and ethnicization but also having to perform a tricky balancing act between the two dominant groups (German-Italian). Migrants generally have to choose between German and Italian to access schools, public bodies, and grants.

With the legally required declaration of language group membership, you can—like people from bilingual families—first select the collection category "other" to declare yourself as one of the three official language groups in the next step to distribute jobs and funds. Beyond the formal act, this also has more subtle effects for bilingual and migrant families: "The wealth of experience of the people in between is therefore not publicly recognized and finds no language. [...] They are pushed into the anonymity of private retreat worlds or are drawn into the pull of the dominant population groups." (Peterlini H.K. 2016b, p. 159). The structural divisions propagate into membership opportunities and identity offers.

On the one hand, this increases the complexity of the processes between adaptation and distinction. It requires an assignment to an (alleged) cultural normality, divided into closed societies. On the other hand, the foil of normalcy to which adaptation is demanded is also more diverse due to this division and perhaps enables more diverse hybrid creative solutions.

Migrant, Black, & Tyrolean Rifleman: Story of Painful and Determined Integration

This system of shared inclusion and exclusion dynamics has experienced a pointed exaggeration, and personal resolution in the case of a young man one-sidedly hyped up in the media. John Christopher Valdez was born in the Dominican Republic in 1994 and came to South Tyrol in 2000 when he was six. After the separation from his father, the mother initially emigrated alone, leaving her son with her grandmother in the Dominican Republic. Then the grandma followed with John Christopher. In the meantime, the mother had found a South Tyrolean life partner who loved the young boy.

On the other hand, John Christopher essentially lost contact with his father, who, meanwhile, had died in the Dominican Republic. In his only detailed interview so far (Peterlini H.K. 2020a, the interview is from 2017), he named the South Tyrolean father alternately in English as *the dad* (in English), *stepdad,* or *Tata* in Tyrolean (ibid, p. 209. It is the story of a hard-won affiliation, with experiences of exclusion and discrimination in middle school after he had easily passed through elementary school in his memory.

> *"In middle school, I started getting the first beatings because of the skin color and so on. Until then, I didn't even know there was anything like racism and that shit. Children do not worry about whether one is black, it is more likely, sorry, for malformations and such, but they do not ask about skin color or origin. They are interested in whether they*

can play.th you, and if that fits, they don't ask, at most once, 'why are you black', but
only out of curiosity, that is not racism,'...]. Then in middle school, I felt it massively
that they ran after me, pus, hed me around, broke everything. I often went to the
teachers, but they couldn't do much. My dad, above all, stood by me. He also told me
that if things go on like this. Hell organize a few people – that's the South Tyrolean
dad who married my mom here, who stood by me." (ibid).

He begins to fight back with the "growth spurt" (ibid). In the narrative of his
attempted assertions "in serious games of competition" (Huxel 2014, p. 69), there
are unmistakable traits of male struggles for superordination and subordination,
as described by Katrin Huxel based on Pierre Bourdieu (2002) and his studies on
"gendered and gendering habitus" (Huxel 2014, p. 69). Domination struggles can
therefore be understood as "instruments of positioning as a man or boy," which,
in conjunction with attributed and/or assumed ethnicity, can at the same time
become orchestration of ethnicity or demarcation (cf. ibid, p. 71) – and therefore
moments of empowerment as well.

"There was a fight, and I suddenly realized what strength I have. Then I actually
started revenge. I looked for the people one by one, the whole second and third middle
school, one after the other. I still missed two of them. I don't know where one of them
is anymore, I met the other one, but a lot of time had passed, he changed, we had a
beer, and he apologized. For me, it is like this, if you have a problem with me and then
have a beer with me ... then it fits." (Peterlini H.K. 2020a., p. 209).

Through sport, John Christopher not only found non-violent ways to test himself
against others but also – for the first time visiting the German school – made
friends with Italian speakers. In sports, the separation of language groups is primarily eliminated; the country's leading football club, FC Südtirol, for example,
is just as multilingual as the top handball club SSV Bozen (Peterlini H.K. 2016c,
p. 292). John Christopher learns to speak Italian fluently, which is hard to do
in school (cf. Abel and Vettori 2017), and receives better grades in the second
language than in German. He accentuates that the Italian circle of friends, in contrast to the German milieu in which he operates, comes from a higher educational
level. "The Italian friends all went to high school. They had a much larger vocab-
ulary – that was not the case in my school. And you only speak High German
at school. Otherwise, you only use a dialect, logically." (Peterlini H.K. 2020a,
p. 206) As an adolescent, John Christopher is in contact with South Tyrolean
language worlds and different socioeconomic milieus. Perhaps the effortlessly
thriving membership in the Italian circle of friends is not burdened by painful
previous experiences, which encourages him to approach German clubs step by
step.

Chance is on his side as well. 2011 – John Christopher was 17 years old at the time – he and his stepfather were sitting in an ice cream parlor when a fire broke out. He bravely takes the fire extinguisher and fights the flames with his stepfather. When the fire brigade arrives, the fire is extinguished, the fire brigade thanks them for the prompt deployment. (Ibid) The next step is a vivid example of how empowerment processes neither originate from a subject that is thought to be autonomous, nor are they only externally determined, but rather take place as a response between phatic impulses and individual response (cf. ibid, p. 321).

So the lived experience with the fire was the (encouraging) impetus for John Christopher to call the volunteer fire brigade commander in his village and apply for the fire brigade. "He said yes, you can come by once. And that's where I went. He later told me, 'I never thought that a black man would come through the door.' But they took me." (Ibid, p. 210) The volunteer fire brigade is an association that substitutes the professional fire brigade throughout the country, except in the cities. As it has emerged from the South Tyrolean tradition, it has long been impossible for Italian-speaking South Tyroleans, and it is still difficult to become an active member. Thanks to his broad South Tyrolean dialect, which identified him on the phone as a "genuine South Tyrolean," John Christopher became the first black firefighter in South Tyrol. The next thing he did was register with the White Cross and voluntary emergency service, which in South Tyrol is on an equal footing with the predominantly Italian Red Cross and has long been considered a German emergency service – here, too, a German-Italian dichotomy.

John Christopher learns to move in the village-like, predominantly German-speaking ethnic groups. He intercepts racist hints: "Negro jokes are sometimes made, but I tell the best Negro jokes myself. It depends on how someone tells a joke. You can feel whether it's hate speech or racism. Then I take someone aside and say that you are going too far for me. Most of them will understand that too." (Ibid, p. 210 f).

Right in the heart center of the South Tyrolean ethnocentric culture, the boldest step is his membership with the riflemen in his place of residence. The inclusion of a black person is a media case, on the one hand, praised crudely as a signal of opening ("skin color of the riflemen is becoming more colorful," *Südtirol News* 2013, *Ein Tirol* 2013). At the same time, it reveals as an expression of hidden racism, if it is a sign of openness to accept someone who differs from other South Tyroleans based on nothing but skin color. On the other hand, the question of whether a dark-skinned man can defend the Tyrolean *Heimat* also triggers open racist spikes. The right-wing patriotic Facebook page *"Ein Tirol"* (One Tyrol) spoke, concerning the cited *Südtirol News* article, of a "loss of identity"

(*Ein Tirol* 2013). The company that had admitted John Christopher had several withdrawals. (Peterlini H.K. 2020a, p. 211).

> "There, I was wiped out by the population, by my own people, from all of Tyrol and, yes, by the Italian media. Media, newspapers, Facebook, I actually had... thoughts of suicide. You can't imagine that. People can be brutal. The company stood by me, except for a few. I simply never show a photo of myself in costume for the sake of the riflemen." (Ibid).

The irritation that occurred when the first and so far only black man was admitted to the riflemen association corresponds to the processes that Bhabha (1996, p. 58) sees in adaptations. These constantly change parts of the society to which migrants adapt or challenge their self-image. By previous media visibility – the mentioned *Südtirol News* article has been removed from the Internet and is only present in the reproduction of the *Ein Tirol* Facebook page – the excitement was quickly calmed down. However, new questions arose just for the discourse about Tyrolean identity nevertheless.

> *"Someone said I am not a genuine Tyrolean. Then I said to him: 'Good, fits. But now I ask you: could you choose where you were born?' He said: 'No.' 'And could you choose your skin color?' He said: 'No.' I said: 'If I could have chosen it, I would have decided to be born here and become white. I couldn't decide that. So you can't judge me by what I can't decide. You can hate me for what I decide, but not for what I can't change.'"* (Ibid, p. 211).

What is interesting about the argument is that, on the one hand, it de-races being Tyrolean, which in turn indicates how much racist ideas are always associated with ethnic or cultural affiliation (cf. Radtke 2013). For Leiprecht (1992, p. 102), ethnicity is often a "language hiding place for a race."

In itself courageous and clever, John Christopher's argumentation has – theoretically seen – also a double-edged side. Behind the statement that he has adapted everything he could, except for the color of his skin, which is difficult to change, there is a commitment to adaptation, which he has successfully achieved. He has done everything he could to adapt. Indeed, if possible, he would ultimately change his skin color or at least wish he was born white. This is comprehensible in John Christopher's personal history and shall not be valued here. But it follows from this on an abstract level that the more or less subtle pressure to assimilation in ethnocentric cultures is faded out. This leads to a lack of comprehension towards those who are not able or willing (which often merges) to adapt so successfully.

"Because quite frankly, if a Muslim would come here now and ask if he could be with the riflemen, I would say 'no' myself. But not because he's black, but because his religion doesn't conform to ours. We are a traditional association built on the Christian faith. If you convert to Christianity, we can talk about it. If you stay Muslim, I'm sorry. You do not belong with the riflemen if you cannot pray with us. I came to the riflemen because I just decided to stand by this country and its tradition, to represent it." (Peterlini H.K. 2020a, 212).

The statement "to stand by this country and its tradition, to represent it" is almost identical to that of the young rifleman Ingo in 1997–2009, based on the same mythical foundation. "My dad has always told me the stories of Tyrol about Andreas Hofer, how he fought for our country, how the French shot him, that they had to shoot twice until he fell over, that moved me." (John Christopher in ibid, p. 212) This self-invention based on narratives and identity offers in the country of arrival goes hand in hand with a – loving but decisive – downgrading of one's home country: "I know little about the Dominican Republic. When I was on vacation there, my biological father always sought and visited me. When he died, I was sad, but I was here at home. My step-tata is my tata. Logically, I still love the country where I come from, but I'm only here at home. In the village, in the mountains, 'I have protection, I'm fine here." (Ibid).

Leeways of Survival From Assimilation to Empowerment

Ethnicization shows itself in the interweaving of foreign and self-ethnicization as a possibility of youthful self-constitution, which can be initiated and strengthened by ethnocentric structures of a society. Young people in ethnocentric communities receive strong role models from myth and history. On the one hand, those allow them rebellion (against the ailing modern society and in recourse to heroic deeds and victim stories from the myth) and at the same time convey security in a conservative world order. In this mixture of traditional revolt and social upheaval, ethnicization also enables a form of becoming a subject that does not have to deal with ambivalences. Ethnicization makes it easier to exclude given multiplicity and fragility in favor of sharply cut self-images and worldviews by suppressing all inconsistencies in life and undesired or undeveloped facets of one's concrete person and/or by projecting them onto them enemy images. In this complex response process, just pointing your finger at the young people, oriented towards offers and differentiation options, is not sufficient from an educational perspective. It is crucial to understand and address the conditions that favor such identity offers – often existential fears, feeling exposed in an increasingly non-solidary society, loss of trust, loss of perspective, threshold experiences before changes with an unknown outcome.

Against this background, it seems understandable that ethnicization serves as a resource for identity formation. It can also cast a spell over young people of migrant origin, whether towards the so-called culture of origin in contrast to the majority society or assimilatory adaptation to it. They take the opportunities they find and respond with the options they have. On his Facebook page, John Christopher also posts photos of car races he participates in, trips to Venice, rescue missions, and – in the time of the "Black lives matter"-protests 2020 – humorous criticism of racism. For example, when a black person wonders why all good things are white and all evil things are black: "The Angel Food cake was the white cake, and the Devil's Food cake was the chocolate cake!".

To be amazed or to find it strange that a young black man in South Tyrol feels at home in the old customs of *Heimat* and pretends to be a patriotic South Tyrolean would be to fall into the trap of racialization of culture. It would mean that one would have to be born here to succumb to the temptation of a strong and ethnoculturally conceived home. Simultaneously, it is necessary to consider that John Christopher is not only a South Tyrolean patriot but there are other facets of his way of life and his idea of identity. Otherwise, we would reduce him to that one facet of identity that we take under criticism. The cage of ethnicization is trapping not only their indweller but easily also who is judging from the outside.

However, the hardships and longings behind the magic German word *Heimat* are neither linguistically, racially, nor culturally justified but have to do with existential issues and living conditions. From a pedagogical perspective, this does not require firm judgments but searching questions.

Daring the Risk of Relation *Heimat*, Identity, Human Image: Perspectives of a Weak Pedagogy

The Difficult Renunciation of Secure Views of World and Man

The examination of nationally colored political identity designs and ethnocentric conceptions of the world is a current challenge for pedagogy. Some years ago, I held intensive "writing workshops" at high schools in South Tyrol. Since I encouraged frank statements, I was often lost afterward about some xenophobic comments. I also felt the perplexity of the class teachers. How do you deal with students expressing right-wing nationalist ideas, claiming racist stereotypes, and reproducing National Socialist slogans?

Conceptions of *Heimat* of young riflemen do not necessarily fall into the categories of right-wing extremism and political radicalism. Still, they are often unconsciously rooted in the same or a related intellectual breeding ground: the exaggeration of one's own, from which everything wrong is cleared in order to project it onto an enemy image, the lifting of one's own good world from a strange, threatening world. "Where do 'desirable *Heimat* ties' end, where does right-wing extremism begin," quoted the South Tyrolean daily newspaper *Dolomiten* (2009, p. 9), Sabina Kasslatter Mur, State Councilor for Culture and Education. Alarming individual cases at the time, currently forced by militant movements such as CasaPound in Italy and the Identitarian Movement in Austria and Germany, had an enormous upswing.

Educational answers to youthful patriotism, which can range from feelings of *Heimat* or awareness of *Heimat* to militant right-wing extremism, require a clarification of the worldview and view of people in pedagogy. A possible starting point for this is a pedagogical attitude, which–as difficult as this is–goes without safe and firm images of man and world in favor of an open exchange between

© The Author(s) 2023
H. K. Peterlini, *Learning Diversity*,
https://doi.org/10.1007/978-3-658-40548-9_5

people. This is a guiding principle for Hannah Arendt: "Every 'idea of man-kind in general' understands plurality as the result of an infinitely variable reproduction of an original model and consequently denies from the outset and implicitly the possibility of action." (Arendt 1960/1999,[1] p. 17).

By fixing a specific image of humankind, human action would "be an unnecessary luxury, a capricious interference with general laws of behavior. [...] Plurality is the condition of human action because we are all the same, that is, human, in such a way that nobody is ever the same as anyone else who ever lived, lives, or will live." (Arendt 1958, p. 8).

It cannot be about raising people's awareness of their *Heimat* or 'healthy' patriotism, nor about therapy for 'correct' thinking, however excellent and democratic it may be, and certainly not about 'depriving' themselves of any ideas. The approach in this work lies in opening up accesses and making it possible to recognize meanings behind the outspoken, which are not judged according to their correctness or falseness but instead make it possible to talk about them. It bases on a narrative pedagogy that seeks a way out of normative educational strategies and moralizing judgments by telling and listening. On a theoretical level, this comes close to the "weak thinking" of the Italian philosopher Gianni Vattimo and finds a coherent starting point in the anti-pedagogical movement around Heinrich Kupffer et al. (2000). The group proposes a pedagogy that does not cling to hierarchical and normative certainties but instead exposes itself to the "risk of a relationship" (Schiedeck 2000). Pedagogical responsible statements "are not the character of expertise that explain causal relationships, but are reflections that open up horizons for questions and necessitate decisions." Pedagogy understood in this way "merely claims to offer 'narrative' forms of explanation and tries to create space for many different narratives" (ibid).

For Vattimo, deep psychological hermeneutics are neither an attempt to examine its objects psychologically nor a renunciation of rational understanding (cf. Vattimo 1997). "Weak thought" (Vattimo 2013) means renunciation of (impossible) metaphysical clarities and encouragement to deal with ambivalences and uncertainties in an entirely rational way. According to Vattimo, hermeneutic philosophy does not establish new metaphysics but "means accompanying being

---

[1] This quotation is not found in this wording in the English original "The human condition" (1958), which was compiled from Arendt's lectures, but only in the German edition "Vita active" (1960/1999), which she translated herself. The quotation is therefore taken from the German edition. The retranslation from German into English proposed above was done by the author. The conclusion about plurality in the following text-passage is also found in the English original and is quoted from there.

along on its twilight journey and preparing for a post-metaphysical world" (Vattimo 2013, p. 51) The plea for weak thought (ibid, pp. 15–24) serves for this work as a plea for a weak *Heimat* that is freed from a "strong" concept and thus becomes more flexible, adaptable, and receptive. A "weak *Heimat*" would rather correspond to the patchwork situations of postmodern and globalized lifestyles and enable *"Heimat"* in them because it can face them instead of resorting to a simulation world. And something similar can be said about the idea of "weak identities," which does not mean a weak person, but a person who gets involved in the ambivalences of life, who does not respond to insecurities with rigid adherence to hardness and rigidly formed traditional patterns of thought, but who–aware of his own ambivalences–gets involved as empathetically and creatively as possible, who accepts life as it is in its unpredictability and diversity.

The idea of "weak thinking" or "weak *Heimat*" carries precisely this risk of misinterpretation. In education, there would be a risk of misunderstanding in the sense of accepting anything, closing your eyes, and looking away. That is exactly not meant here. Weak thinking, the open approach to many opportunities to see, experience, believe, design demands responsible and equal subjects in pedagogy, who are merely aware of the relativity of their terms and worldviews but do not escape from them arbitrarily. It is precise because the Other, in the sense of Arendt, is recognized on the one hand as being equivalent but is accepted in its fundamentally different nature, which makes confrontation and exchange possible.

A Phenomenological Exercise: Reading Traces on a Drawing Instead of Diagnostic Judgements

The here proposed pedagogical attitude fits in with the phenomenological approach (see the following chapters), which is to bracket and hold back the judgment or the pedagogical-psychological diagnosis in the educational process as far as possible in favor of an attentive looking and perceiving what is actually said in which words, with which gestures, with which bodily language and in which contexts, in order to search for answers from this concrete perception of the concrete other. Or even better: not to search for answers at all, but ask questions about how to answer this situation most adequately and feel how the person searches for answers when given time and space.

In pedagogical discourses, the query of perception often ranks behind the pretension of understanding. Perception is simply skipped in favor of classification into categories of understanding. For this reason, in one of my seminars, I let students present pedagogical problems only with a drawing without any additional knowledge about the child or adolescent. The instruments with which pedagogical situations are generally classified are thus not available. There is only one drawing, perhaps a sun and a cloud over two people laughing or with the corners of their mouths drawn down, a fist, rays, lightning, a river maybe. The drawer holds himself back at first. The exercise is for the seminar group to interpret the picture, not in terms of what might be behind it, but what is shown in the image: the colors the scene is painted, thicker or thinner strokes, symbols. Gradually this perception turns into interpretation, into thoughts about what this or that could mean. The obstacle of looking closely before interpreting creates a variety of possible meanings. After the exercise, the person who painted the picture comes out and offers the context. The interpretation becomes narrow again, but in a more multi-faceted way. The person drawing has received offers of understanding which partly contradict their interpretation, partly reinforce it, and partly make visible additional aspects that would otherwise have escaped through the hasty judgment. (cf. Peterlini H.K. 2020b, p. 8 f.).

Perception as a pedagogical need means precisely that: Taking a breath before the quick interpretation, avoiding abbreviations of classification by looking, listening, and feeling. If we know from a child that they have been diagnosed with ADHD, it will be difficult to perceive them outside this behavioral model. It is subjected to the power of that diagnosis which can help take measures for the child but may also hide a lot of what this child is too, can be, and wants to be beyond the diagnosis. The child will be misunderstood in its surpluses and plurality, and much will not be done to help the child escape its diagnosis.

The phenomenological approach of avoiding statements about what things are supposed to be, instead of dealing more intensively with how they appear to us in concrete practice, is a way out of narrow-minded pedagogical judgment. In this attitude, we open our eyes, ears, and compassion for extended meanings of what we encounter and challenge in our pedagogical work. It expands the grids of classification of what social or personal behavior may mean by preceding the quick judgment with the delaying exercise of perception. And it also raises the possibilities for action, for searching questions in the concrete pedagogical work.

So the phenomenological practice leads on and joins a "weak pedagogy" as described above. The narrative approach involves the pedagogical subjects in a participatory way, not judging them but exploring their situation and perception and giving them spaces for speaking and acting. The relevance for dealing with national identity drafts, with concepts such as *Heimat* and ethnic identity, lies in giving preference to the perception of concrete situations and practices in the world of life and speaking with those affected over judging: namely, to look at their narrations behind the phrases of the discourse, to look at their practices and social actions, to feel into what the motives behind the gesture can be, even if these gestures are unpleasant.

Pedagogical staff at all levels suffer the temptation to cope with irritating phenomena of ethnicization or even racism by directives about the *correct* political identity. The danger of such a strategy is to drive the devil out of the Beelzebub and do just what you pretend to prevent. Kupffer sees a continuum from fascism to now in german pedagogy, namely the tendency "to respond to specific social, political and factual questions not adequately at the respective level, but in general, with an image of man. This image of man itself remained essentially unchanged" (Kupffer 1984, p. 9).

The harsh verdict is an appeal not to reproduce the same stereotypes and behaviors in one's method and way of thinking that pedagogues try to overcome. With this, Kupffer ties in with Horkheimer and Adorno, "that even progressive approaches often get stuck and lead to a new dogmatism. The enlightenment once achieved is viewed as an available achievement and then becomes no less repressive than the old state against which it had protested" (ibid, p. 150; cf. Horkheimer and Adorno 1944/2002). Pedagogy was particularly motivated by the demand that Nazism never emerges in Germany again, which made it fall, for Kupffer, into the trap of hidden authoritarianism: "The 'dialectic of the Enlightenment' leads to new stubbornness." (Kupffer 1984, p. 150) Following Adorno, Kupffer warns against the misconception of fighting fascism by treating the superficially responsible "authoritarian personality" and leaving unchanged social conditions under which it arises (ibid, p. 5).

For Dieter Sinhart-Pallin and Martin Stahlmann, "educating for Democracy," which might appear to be a pedagogical task, "is a contradiction in itself because the intention already presupposes a hierarchical view of the world. Behind this view is a reduced relationship between education in the form of a binary 'if–then' model and an image of human beings, which the child regards as 'unfinished' and as a manipulable being without its own justified will" (Sinhart-Pallin and Stahlmann 2000, p. 10). The chance for pedagogy lies in abstaining the right image of man and instead relying on an open exchange about concepts of self and others, as Jürgen Schiedeck encourages with reference to Martin Buber: "In Buber's sense, what happens has the character of a gift." (Schiedeck 2000, p. 70).

Enduring Ambivalences as Educational Challenge

The discussed approach is not a trivial exercise. Sometimes it asks for enduring almost hurtful things because it is evident that pedagogy takes the floor against injustice, against exclusion, against contempt for the weaker. The humanistic impetus lingers from Goethe to the present day: "Nobel be man, helpful and good" (Goethe 1789). Ultimately, not detaching yourself from such a pedagogical concept means excluding everything that does not fit into the picture and devaluing and removing it in the same way as in racist and nationalist ideologies: to design a "right" person by resecting the inappropriate, the inconsistent, the contradictory, the aggressive, the violent. In this sense, Thea Bauriedl feels desperate about the lack of progress by man-kind despite the Enlightenment, the 1968 Revolution, and educational reforms: "Then as now, we cannot stop waging war against the environment, against the Third World, against the weak." (Bauriedl 1988, p. 201) Even though written 30 years ago, it is more topical than ever. Bauriedl unfolds the assumption that the undigested residues of inadequate or absent mourning continue to work behind the repetition compulsion. When it freezes in depression, the repressed and oppressed demand for acting out: "The basic feeling of depression is despair because contact with fellow human beings and with reality is lost. The result is a desperate punching around, the inability to stop destroying yourself and others. Desperation does not always have to be experienced consciously. It can also hide in optimistic gestures of power that then drive others to despair." (Ibid) A key to pedagogical answers can almost only be to make that what can lie behind "striking around," behind gestures of power and threats, behind the constructs of exclusive designs of *Heimat* and identity, communicatively and emotionally accessible again.

"Remember, repeat and work through," but in small steps, is the only possible path for Bauriedl (ibid, p. 205). The pedagogical index finger that one should not say this or that or that it is entirely wrong closes all access to the repressed. While transferring Bauriedl's psychoanalytic considerations to pedagogy, the goal cannot be "that the oppressed so far are finally coming to power, that is, that the power relations are only reversed" (ibid, p. 208). The South Tyrolean example of the Germanization of Ladin place names by a German majority, even while complaining about the Italianization of their German names, shows how the reversal of rule relationships does not entail a change in rule methods.

The attempt of instruction and conviction to "right" thinking touches on the sensitive psychic construct of victory or defeat. Suppose the pedagogical intent is to drive out the undesirable. That case strengthens the holding on to it precisely because letting go would mean failure to the educator's authority, the adult, the political mainstream. According to Gianni Vattimo's "weak thinking," pedagogy of the weak worldview would have to forego the certainty of their own strong position and the safe knowledge of the educational institutions. A weak pedagogy is not weak but weakens itself to enable exchange with the weaker in the first place and develop the ability to understand in the first place, wherein a strong position the power of judgment is lurking.

One more thought: The insight that pedagogy cannot solve everything, but that it has to get involved in every new, unpredictable and uncontrollable relationship process also relieves the frequently encountered phenomenon of excessive demands on educators, both parents, and professionals. The assumption that education can 'repair' everything leads to frustration and powerlessness at the latest when this claim is missed: if the child or the adolescent do not function as they would have been brought up or taught. The idea of a weak pedagogy frees from this pressure to succeed. The success or failure of educational processes is no longer solely the responsibility of the child or adolescent, who can no longer be held exclusively responsible for their failure. However, success or failure is not only the educator's responsibility since the educator alone cannot achieve the educational goal. Both always fail, just as success belongs to both sides.

Here, too, a dichotomy that has established terrible hierarchies in the pedagogical context is put into relativity again. Learning is a shared experience. Teachers can gain a lot from learners if they descend from the hierarchy ladder and do not believe they can determine everything, know everything, have everything in their hands. It is often unclear what success or failure in educational processes means. Can there not also be a failure in success and success in failure? To doubt causalities in educational methods implies, on the one hand, mistrusting the functioning of tools, but on the other hand, also trusting that what feels like a failure today can lead to success tomorrow in a roundabout way. Sometimes what has just been euphorically celebrated is a stumbling block on future paths. Sometimes educational interventions, as misguided as they may seem at the moment, are belatedly blossoming like flowers. Ultimately, those responsible for education, whether as professionals or as parents, can only risk themselves anew by getting involved with the people entrusted to them. Being authentic here does not always solve all problems, but it is often the only thing we can do. And it signals to the child, the young people, but even adults: I am here, and I get involved with you, do it the other way too, maybe we will get a bit further.

"Weak thinking" also means going to the level of lifeworld and practice in dealing with worldviews and identity concepts. Only at this level would it be possible to perceive and subsequently better understand what is no longer accessible at the system level, where it can only be accepted or rejected as a fixed pattern. There is no need to accept or reject at the level of everyday life and practice exploration. Still, it will be possible to communicate what the system constructs are based on: What is meant when *Heimat* is understood in one way or another when enemy images or projections are held against any rationality. "Weak thinking" is also the only way to give the other person the chance to let their own thinking weaken, to open themselves to what the other is saying, without the fear of being subjugated by their strong thinking. "Weak thinking" thus avoids the trap of defeat, which we fear and which leads to hardening, both on the side of the educators and their subjects.

Narration as Way of Exchange Experiences

A narrative approach in pedagogy requires reflection on communication. Speaking with Bauriedl, you need to be prepared to risk and accept the "defeat" in a conflict (Bauriedl 1988, 208 ff) because otherwise, all communication strives for victory, and being right at all costs becomes a must and inevitably strategic. This pedagogical style aims to weaken rigid identity designs without necessarily changing or renovating them but only making them communicatively and emotionally accessible. It makes a difference whether someone says that they hate foreigners, even though someone argues that foreigners are good people, or whether they say that they hate foreigners, but also reflect about good experiences with some foreigners. Consequently, confronting a narrative setting how someone else interprets one's designs for *Heimat* and identity, what reflections they bring about in the other could be a further step for learning in the sense of increasing awareness through confrontation.

For Duccio Demetrio, the meaning of telling one's own stories lies in the fact that "memory becomes present" (Demetrio 1998, p. 24) The experience that someone else is interested in their own life-stories increases their appreciation of themselves and others (Anzaldi et al. 1999, p. 29 ff.). It offers opportunities for a new approach to their own life issues. Narrating is always restructuring what happened, looking for and finding a chance to deal with the past differently.

The young riflemen and marketers involved in the presented research project remembered the actual contents of the conversation from 1997 surprisingly well, even twelve years later. Some spontaneously said that from the first encounter on, they would have thought about *"Heimat"* again and again. In the sense of narrative pedagogy, this would mean that telling about *Heimat* has set something in motion–retroactively and with a forward-looking effect (also in the sense of retention and anticipation). At certain life moments, the interviews' memory mainly was associated with a review of whether this was still correct or had to be adjusted. Simply speaking about *Heimat* in dialogue with others would have changed the idea of *Heimat* and opened it up a bit for revisions, adjustments, extensions. These would then be "small stories," personal deconstructions that created the prerequisites for later new deconstructions to deal with the lifeworld.

These experiences give hope, even if the small (weak) stories come up slowly, only occasionally, and mostly only temporarily against the prevailing prominent (strong) narratives, that our world would be safe if only nothing new and strange came in from outside. The return of South Tyrol to revanchist demands, to new xenophobia in the sense of a shift in enemy images (from the Italians and Germans to migrants), to a nationally charged *Heimat* boom in recent years shows that this myth has lost none of its strength. On the contrary, the attitude towards migrants shows that German and Italian right-wing parties have congruent positions but are in no way reconciled in a common new enemy. They agree on who to hit but are immediately hostile to each other regarding their respective ethnocultural interests. For instance, while German and Italian populists in South Tyrol agree that migrants should not overwhelm their schools, they also argue whether Italians should be allowed into German schools. A school for all does not even come to mind.

The enemy image was, strictly speaking, not shifted but expanded by a category. The contradiction that German and Italian right-wing populists, on the one hand, accuse each other of lacking understanding of the other but struggle separately with the issue of migrants is not even evident to the actors. Behind the protective masks of their nationally rigid identity, the fencers cannot see that they are doing what they complain about to others. If German-speaking and Italian-speaking South Tyroleans had learned in working through their conflicts to deal with the other as *Other,* then they would also have learned how to deal with migrants.

Modesty in the claim to healing is imperative. No final victory is conceivable in the confrontation with myths and worldviews occupied by the enemy because the destruction of the myths almost inevitably heralds their resurrection. Ultimately, destroying myths means resorting to the hammer of "strong thinking," which knows better than the myth and takes its place as a new myth of truth when it shatters the myth. Ideologies cannot be combated through ideologies but require discussion through deconstruction, disassembling in the knowledge that something else is being assembled that must be entrusted to further disassembly.

In this perspective, Aleida Assmann refined the memory theory by rethinking the concept of social memory. In contrast to collective memory, this requires conversation and social exchange between group members. It is more transient and remains bound to the discourse among the living. While the collective memory is reduced to narrow traces of memory, can solidify over the centuries, and tends to simplify, the social memory is open and opening. The stories of young adolescents about *Heimat,* the stories about them are moments in time. They are tied to a temporary dialogue, nor can they be poured into clear laws and formulas.

Therefore, the aim of the approach proposed here is only the wish that it may inspire pedagogy in its diverse fields, in school classes, associations, youth groups, and patriotic movements, in politics, culture, society. We should try to create moments of social memory as small stories that are temporarily opposed to the collective's monumental memory and weaken it in the sense of more serene, opened, expanded approaches. In the following second section of the book, this will be taken up again and deepened explicitly in different pedagogical perspectives, which not only currently represent a significant challenge for dealing with diversity: Concerning the divisions and splits in education along categories of difference such as normal/disabled, talented/untalented, gender, culture, language, origin, about the challenge of ethnocultural and racial discrimination in society and institutions, and finally to the human handling of nature that we think disconnected from us. Diversity learning means a refined perception of plurality, on the one hand as the various conditions of normality, our being how we are and our being in the world, which makes us unique, and on the other hand as a target for discrimination and devaluation due to a narrow understanding of normality. It is not an easy field, the pedagogical one, but we must dare it.

The (dis-)order of Experience—Facing the Splitting of Learning and Living

The Split School. *From Selective Normality Concepts to a Phenomenology of Diversity*

> *"And everyone who doesn't adapt*
> *Is declared a problem child*
> *And anyone who is too lively*
> *Gets a pill, so she doesn't bother anyone*
> *And with that, you cheat yourself because*
> *No child is a problem."*
>
> Sarah Lesch, song "Testament," written 2016

The hidden curriculum of the selective school

"Inclusion is good, but how do you get a homogeneous class together?" The question, or rather the sigh, of a student as part of the course "School socialization and social inequality" (at the University of Klagenfurt in 2018) condensed in her concern for well teaching several misunderstandings about inclusion. The first thing that catches the eye is the idea that the goal of inclusive lesson design can be "a homogeneous class," that is, the heterogeneity that students bring into a school class would have to be standardized to homogeneity, through whichever educational and didactic efforts it takes. Between this supposed task and the concern about *how* (not *whether*) this can be done, another misunderstanding is the belief in the didactic and pedagogical feasibility of learning, a causality assumption that all learning, including inclusive social learning, can be accomplished through suitable teaching methods. Teaching reaches a limit where it has to *achieve* something that simply cannot be carried out. And over the whole sentence is an – unspoken – understanding of normality, into which all deviations in the class would have to be fitted through homogenization – the ideal child, which does not exist.

© The Author(s) 2023
H. K. Peterlini, *Learning Diversity*,
https://doi.org/10.1007/978-3-658-40548-9_6

The shadow of this deep-seated claim as a *hidden curriculum* (Sambell and McDowell 1998, p. 391) of *getting* "a homogeneous class" *together* by inclusion is the implicit idea that such a class could be easier if only the proper selection were made. In the seminar mentioned, this led to a mind game: Take a school class from a hypothetical unit middle school that has been in existence in Italy for over 40 years due to the non-segregating upbringing there (cf. Peterlini J. 2015). Then one begins to separate different categories of children or adolescents one after the other to see whether the result is a homogeneous class: for example, first omit all children with disabilities,[1] because their stereotypical differences are the first thing that catches the eye; students follow who would previously have been described as mentally disabled; in the next step, talent grids could be introduced to sort out children with learning difficulties and those with top talents, who could then be transferred to other school settings – support classes and high-performance classes; the grid would very quickly call for a refinement, since skills if these can be easily determined, vary greatly depending on the subjects and disciplines. In a hypothetical homogeneous class, the computing genius, whose German essays are somewhat brittle, could hardly be taught with the writing artist, the sports ace hardly with the slightly less sporty intelligent nerd. Forgive the stereotyping used to intensify the project to form a homogeneous class through selection. Finally, it could come to mind to sort the children by social or regional origin since these factors also generally stand in the way of homogeneous class formation. The Aha-moment in the seminar group quickly evidenced that there would be no child left. The homogeneous class would be a class without children.

Despite the palpable absurdity, the idea of homogeneity as a school goal to be achieved is a powerful discourse, partly hidden, partly unhidden. It is inscribed both in the doubts whether an inclusive school is possible and in the efforts for inclusion. A recent example is the introduction of language support classes in Austria for students who "have no or only minimal knowledge of the German language" (BMBWF 2018; *Kurier* 2018). Such a measure may have language-didactic advantages and disadvantages. Beyond didactic considerations appears the illusion that it is easier to learn and teach in homogeneous classes. These perfect classes would be possible when all students with inadequate German knowledge receive special lessons, and all *normal* students are prepared without the linguistically weaker students in *purely German* courses. Such fantasies emerge in the very first reaction of a reader to a media report (*Standard* 2014), which has criticized the selection of children

---

[1] In German, the term "disability" (*Behinderung*) is highly problematized and used instead of the preferred term "impairment" (*Beeinträchtigung*) only to make the underlying social discrimination evident.

of migrant origin. The reader replied: "All my friends with family moved to the surrounding area in Lower Austria. The homogeneous classes form on their own." (Ibid).

Here, too, is the idea that the migrant classes left behind by the departure of the German families are indeed homogeneous in the negative sense, and the more peripheric Lower Austrian schools in the positive mind. Such imaginations refer to the narrowing of homogeneity concepts to a single characteristic of deviation, namely the one that first catches the eye. The current 'deviation' is the so-called *migration background* as a unifying negative characteristic. We see the problem of this term: it is ultimately not meaningful from a social science perspective (cf. Hamburger 2010, p. 17) because it assumes that the children of the migrant origin or family history are all equal. For the supposedly positive homogeneity of pure German Classes, the unifying characteristic is that they are not migrants (cf. Huxel 2014, p. 71). That this does not mean homogeneity here or there reveals itself only at that "second glance," which Niklas Luhmann describes as the beginning of scientific questioning, with which "new questions and concerns come up" (Luhmann 1981, p. 170). These questions and concerns could be:

- Is it conceivable that all children have the same language skills in school classes not influenced by migration? And would someone for children with language difficulties from—say—Lower Austrian families demand a language support class?
- Is it conceivable that if all children and adolescents who were not linguistically adequately competent were separated, the classes homogenized in this way would be really homogeneous, for example, in terms of educational, family, social, economic differences, but also terms of their competence in different subject areas, including language?
- Is it certain that children who are excluded from the classes by a language test would not catch up and flourish very quickly in these classes, while they may be "kinked through the separation and forced support before they can rise" (Müller 2018, p. 10)?
- Does a school policy program that focuses on perfect monolingualism—not even reachable for domestic children—correspond to the educational policy requirements in a migration society? Or does such a program cling to the monolingual habitus (Gogolin 2008) that shapes education in the nation-states, even if they have long been multilingual in a globalized society and economy?
- Shouldn't schools instead be designed to value the existing wealth of multilingualism as a special competence (cf. Gnutzmann, 2004, pp. 45–47) and to anchor

it accordingly in the curriculum (Gogolin 2008, p. 23), instead of continuing to devalue and reject it as a disturbing deviation from the desired homogeneity?
- And finally: Who can seriously say that learning is better in homogeneous classes than in heterogeneous ones?

Inclusive Openings and New Closings

It is no coincidence that the introduction of language support classes in Austria 2017 goes hand in hand with the gradual opening of the concept of inclusion in international discourse. There have been significant steps since the long efforts of the disability movement from the 1960s on and boosted by UNESCO's demand (2009), that all people "regardless of their differences" like special learning needs, gender, social and economic requirements must have equal participation in education. The understanding of inclusion over the past decade has been gradually extended based on the UN Convention on the Rights of Persons with Disabilities (United Nations 2016). It is no longer just a question of how children and adolescents with disabilities can be integrated into regular schools, but how schools and lessons have "to do justice to the diversity of the requirements and needs of all users" (Biewer 2010, p. 193).

The Austrian example shows that this conceptual opening to diversity aspects that can no longer be defined a priori and by clinical diagnosis reinforces rather than reduces institutional resistance to inclusion. While the conventional understanding of inclusion or integration was about coordinating regulations and measures and training and further education for certain identifiable groups, schools and school policy are now challenged and partly overwhelmed by an understanding of inclusion that affects everyone. "Preservation and strengthening of special education" and "Reintroduction of special education" (Bundeskanzleramt 2017; see FPÖ 2017, p. 62)[2] are only the particularly salient projects of the ÖVP-FPÖ coalition, which point to at least the desired return to a selective school system.

---

[2] The program of the ÖVP-FPÖ government was intended for the period 2017–2022 and was suspended with the fall of the government in 2019 after the Ibiza affair around Vice-Chancellor Heinz-Christian Strache; the following ÖVP-Grüne government retained the German support classes, the before Minister Faßmann (ÖVP) went back in his position till 2022. Thus the government program 2017–2022 was removed from the website of the Federal Chancellery; in 2020, it was still available on the FPÖ website (FPÖ 2017).

At first glance, a diversity-conscious attitude seems to be behind the measures: "Because every student in this country has different talents and skills, has special interests and may need support and catching up in certain areas." (Bundeskanzleramt.2017, see FPÖ 2017, p. 59) These "differentiated requirements" should now be met "through a differentiated structure of school types" (ibid). The language support classes show how quickly specific support can create selection and exclusion through separate offers. The elimination of children who, for whatever reason, have difficulties in terms of language consequently carries on the already present (cf. Feyerer and Holzinger 2017) paradigm of special education selection by transferring it to new groups. The children from families with a migration history are the first mark that others could follow. Suppose children can be kept away from regular lessons for linguistic reasons. In that case, it consistently should not be an issue to separate those who show poor mathematic performance or have other learning difficulties. The path from special education to inclusive school, as it started from the Salamanca Declaration Steps taken in 1994, would at least be interrupted, if not entirely abandoned.

Concerned that homogeneity in a school cannot be achieved for everyone, the attempt—previously fictitiously played out and doomed to fail—would begin to restore homogeneity through group and category formation in a selective school system. This would be, of course, only possible for large (inhomogeneous) and dichotomously constructed groups—i.e., disabled people vs. non-disabled, those with learning difficulties vs. normal, normal vs. talented, migration background vs. locals. Although homogeneous learning groups would not be achieved, the attempt would create a new basis for institutional discrimination.

The laboriously achieved international consensus that schools should not be exclusive but as inclusive as possible is being tested anew in Austria by its extension to diversified lines of difference. The requirement that education should "reject labels and classifications" (Biewer 2010, p. 193) in order to "minimize discrimination and maximize social participation" for all children and adolescents (Werning 2016, p. 229) seems to arouse old resistance and create new rejection tendencies. With the labels and categorizations to be abandoned, (alleged) certainties are lost, which constitute the previous and still dominant self-image of school, education, and learning.

Normality Assumptions as the Subject of a Conceptual "Polemic"

At the center of the unsettled self-image of how school should be or should develop is nothing less than the question of normality. For a long time— albeit mostly linguistically veiled—this has determined the distinction between *disabled* and *normal*—the regular school for the normal, the special school for the non-normal. An inclusive approach, which attacks this dichotomy by ultimately multiplying the distinctions indefinitely, inevitably feeds restoration requests according to the lost order. If "everyone is different different" (Arens and Mecheril 2010, p. 11) or "different various" (Lutz and Wenning 2001), then—radically pointed—the idea of what is normal and what is not is lost. Getting involved in this requires a new understanding of normality that does not emanate from children and adolescents as hypothetical ideal products of edu-cational processes but from children and adolescents as they are and will be. School's traditional, deeply rooted normalization mandate, namely to adapt chil-dren and adolescents according to normatively set societal wishes (cf. Foucault 1995, cf. Sohn and Mehrtens 1999), would thus result in the central objective with its potent guiding instrument. The goal, of course, is normality imagined as homogeneous; the device is the dichotomous constructed notion of normal-ity. Attempts to restore the old order through new selections show compensation tendencies against the feared loss of orientation.

Pedagogy as the reflection science of educational, socializing, and learning processes in school, family, and society faces a dilemma inherent to it from its beginnings again. Although education was meant to be an offer for everyone since the Renaissance and again emphasized with the pedagogy of the Enlightenment, it has become an institution for the preparation of social demands on the subject and its economic usability (cf. Ribolits 1997, pp. 19–25). All educational efforts' more or less explicit goal was the functioning person, measured by norms, which could only be brought about by distinguishing the normal from the non-normal. In his unbroken timely work on "The conquest of the child through science," Peter Gstettner writes about the production of normality: "This dynamic is always borne by a polemic that imposes 'the normal' on man in order to accuse him of 'the abnormal' (as negative)." (Gstettner 1981, p. 66).

An essential critique on a simple understanding of normality is George Can-guilhem's analysis of the act of definition with which normality is brought about:

"A norm, or rule, is what can be used to right, to square, to straighten. To set a norm (normer), to normalize, is to impose a requirement on an existence, a given whose variety, disparity, with regard to the requirement, present themselves as a hostile, even more than an unknown, indeterminant. It is, in effect, a polemical concept which negatively qualifies the sector of the given which does not enter into its extension while it depends on its comprehension. The concept of right, depending on whether it is a matter of geometry, morality or technology, qualifies what offers resistance to its application of twisted, crooked or awkward." (Canguilhem 1991, p. 239)

What is remarkable about a normalization act understood in this way is that normality becomes abstract. Not the everyday life of the child and its concrete existence were (and are) "empirical basis for the 'ideal typical' and 'normal' in the development of children and adolescents (and their résumés)" (Gstettner 1981, p. 81), but ideas of naturalness and purity, "i.e., a picture that comes about from the wish and repression notion of the adults" (ibid). Such normalization can be interpreted psychoanalytically as repression from socially undesirable everything (Erdheim 1984, p. 25). According to Gstettner, these tendencies have to be contrasted with a research attitude that understands normality as "what happens every day; every-daily is therefore also the process in which something is called 'abnormal'" (Gstettner 1981, p. 82). Science has the task of seeing through and criticizing the production of normality in its everyday life by understanding "'abnormal' not as something per se excluded, as a malformity given by natural law, but as a definition brought about by interacting people" (ibid). "Phenomena are therefore only constituted as 'abnormal' if they are perceived and/or represented as such by interacting persons." (Ibid).

Gstettner bases this research stance on the ethnomethodology founded by Harold Garfinkel (Garfinkel 1967). The continuously generated assumptions about the world ("ongoing accomplishment," ibid, p. 1) require an equally continuous review of secured knowledge, everyday practices, and conceptual categories – and thus also assumptions about the normal and the non-normal. It is only through alienation of the fundamental beliefs about the world that they become accessible for a critical discussion, namely how members create and construct the assumptions about the world. In developing ethnomethodology, Garfinkel was guided, among other theories, by the sociological implementation of Husserl's phenomenology by Alfred Schütz (1967, cf. Eberle 2009, p. 97). Common to the approaches is the ἐποχή, epoché, the bracketing or withholding of assumptions and presumption, as Edmund Husserl postulated as the starting steps of his phenomenology research attitude as abstention in judgment as far as possible (Husserl 1913/1990, among others pp. 6, 46, 42). This does not mean denying the reality of the world and things but abstaining from ultimate knowledge instead of looking and listening to how phenomena show themselves to human perception and consciousness.

In a critical examination of constructs of normality and deviation, such an approach can be made fruitful both at the level of scientific inquiry and praxeological implementation. This requires pedagogical offers for precisely those overstrain that the dissolution of right-wrong, normal-non-normal categories has brought with it. It is also a matter of weakening epistemic violence exerted on learners by predetermined and thus determining knowledge (cf. Brunner 2021). If being different is no longer defined from the outset through attributable categories, ideas of normality also elude a priori definition. The goal of learning and education is no longer a normative, normalized dream reality, to which deviations have to be adjusted or selected. But: the given diversity itself becomes normal. Normalization means that deviations do not have to be diagnosed, treated, or sorted out, but the "as is" of children and adolescents is perceived as the given and tangible reality. Their "as is" should not be subjected to any delimitation from imagined normality because it is one of many ways of being within an extended concept of normality.

This understanding of normalization requires pedagogical perceptions of difference rather than clinical diagnoses. The dominant diagnostic methods tend to subordinate the required open perception to diagnostic acuity. This dynamic can be seen in the conceptual and operational merging of "support diagnosis" and "selection diagnosis" (SQA 2017, p. 6, table), as operationalized in Austria in a joint initiative of "School Quality General Education Sect. 1" and the Federal Ministry of Education and Science (BMBWF). The pedagogical perspective on normalization proposed here is not limited to selection or individualized support measures, which are still necessary depending on the problem. The disadvantages of exclusion and discrimination practices have to be minimized (cf. Dietrich 2017, p. 125). Instead, it considers the given and phenomenological evident reality of the life of concrete people as normal to give space and attention to this difference based on this prerequisite—in the sense of a school that is sensitive to differences. Equality and difference form the tension between the same normality, provided that being different is recognized as constitutive for an "inclusive normality" (ibid).

Phenomenological Receptivity as a Perspective for an Inclusive Understanding of Learning

How schools can deal with heterogeneity, if this is viewed not as an *impairment* of a desirable homogeneity but as wealth—according to a much-used and little-heeded formula—is the task of different approaches. It includes educational theory and practice, school and educational policy, management, didactics, teacher training, and continuing education. A common worthwhile starting point could be the rehabilitation of one of the two basic human attitudes that Wilhelm von Humboldt's placed at the center of his considerations about people and education (cf. Humboldt 1783/1980, cited from Meyer-Drawe (2018, p. 38). The two principles, necessarily interacting but divided in current practice, are *self-activity* and *receptivity*. While self-activity under different names (i.e., currently the concept of competence) has played a dominant role, receptivity to the world and its phenomena has taken a back seat. Two basic attitudes compete, which emphasize different—but interdependent—aspects of learning: on the one hand, the mastery of the world, its objects, and concepts as self-activity, which expresses itself scholastically as the acquisition of competence, as mastery of canonical knowledge and its usefulness; on the other hand an unsecured openness to the world, as it can always appear new, unpredictable, not canonized, surprising, withdrawn from norms and defined ideas if this—in the attitude of receptivity—is allowed and perceived. Husserl's epoché is—as bracketing of familiar assumptions about the world, in order to be open to new understandings (cf. ibid)—the prerequisite for this, a pause that enables a look, listen, and feel for diversity and the constitutive differences between the world and others.

The following is a Vignette from the research project "Personal, educational processes in heterogeneous learning groups," in which teaching as a crowded inquiry and communication is blind to the need to pause:

Some students have forgotten tasks. Ms. Fassneider asks for ideas on what could be done to avoid this. The pupils talk about how they plan the tasks, the teacher intervenes with a "good," "great" after each word to ask for "further suggestions" without taking a breath. It's Franz's turn: "To do the tasks on the day they are given." "Great," calls the teacher, "who can do it? All hands up!" Florian hesitates, "But there are so many calculations in math that you can only do one every day." Ms. Fassneider: "Right, you must practice every day!" With a significant gesture, she exclaims: "Who else?" She barely leaves a second between the students' responses and their comments, immediately calling up other students: "Franziska!", "Fritz!" Freino wants to explain why he did not do the task: "I forgot the task because I was out" (in an accented South Tyrolean dialect). Ms. Fassneider: "Freino, how do you

talk to me?" Freino repeats (again in dialect): "I forgot ..." The teacher: "How do
you talk to me?" Freino understands and changes the standard language: "I have
not forgotten the task I was missing." "Fine," says Ms. Fassneider and seamlessly
picks up on the material for the current teaching unit: "What is an adverb?" she calls
out to the group. Several children raise their hands. Ms. Fassneider abruptly turns to
Fiona, who is playing with the scissors: "Fiona, listen because tomorrow is a test.
That's exactly what is coming!" Fiona straightens up, puts the scissors away, but soon
pushes the stylus box back and forth, takes out the sharpener, looks into the narrow
opening, and looks at Ms. Fassneider. The latter continues to question the class in
staccato. Fiona quickly raises her hand to ask a question but doesn't get picked. She
reaches under her bench and pulls out the timetable. Ms. Fassneider continues to throw
question after question into the room at the same pace, hardly ever letting those pupils
speak up who raise their hands, but surprised with her shouts: "Flocky!", "Frida!",
"Fiaba!", "Ferdinand!" – After each answer, there is promptly a new question. Fiona
holds up her hand again. This time it's her turn: "Yes, Fiona, now you!" Fiona begins
hesitantly: "Uh ..." ... Ms. Fassneider barks: "Not uh ... go!" Fiona swallows briefly,
then says quietly: "Professor, when do we go to the library?" Ms. Fassneider looks
at her disconcerted for a few moments, and the class is as quiet as a mouse, then she
says in a stern tone: "Do you think this is conducive to learning if you say something
in the middle of the material that has nothing to do with it? We'll talk about that at the
end of the lesson." Fiona looks at the bench. Ms. Fassneider calls Fabian. (B0_FS03,
cf. Peterlini H.K. 2019a, p. 51)

The Vignette shows an example of the pace of knowledge transfer at school,
which leads to sheer breathlessness despite the teacher's interactive approaches.
School and all those involved are under pressure from functional skills development, which leaves little time and space for pause and digression. If canonized
knowledge has to be worked through according to excessive curricula and annual
programs, school misses the leisure that was the inspiration for its ancient pre-
decessor model, namely the ancient Greek σχολή for "leisure" (Liddell and
Scott 1950; Duden 2007, p. 742). This also slowed down and, in some cases,
turned it into its opposite competency-oriented teaching following the European Qualifications Framework (cf. EQF 2008[3]), which, according to its original
intention, wanted to place understanding above the accumulation of knowledge.
The inversion of the goals, in reality, is shown exemplarily in a pilot study on
the implementation of the new framework guidelines for competence-oriented
teaching at secondary schools in South Tyrol:

---

[3] The Framework was edited 2008 and revised in 2017. The quoted study concerns the
version 2008.

"So with us, they put a lot of value on it, the students have to be able to do this and that [...], so with us they really put a lot into it! We could already train 20 professions!" (LF1/B/1).

"For my subject, I have to say that everything is so detailed and precisely specified. There are skills and knowledge, and we can add content, whether we take Goethe or Schiller, Dürrenmatt or Frisch. But how much of something and when I do it and what I want to do is no longer up to me because the curriculum is compact and full. I can only see what I leave out because I have to do everything completely. The framework guidelines are so extensive that I can hardly say that I do more or less there. I have to do the minimum almost everywhere. Otherwise, I won't get there in terms of time and content." (LF1/F/5)

For teachers to engage in diversity that does not fit into normative norms, a fundamental change of perspective about learning itself is required. The—partly overt, partly unconscious—striving for homogenization feeds on an understanding of learning based on the result to reach everyone at the same time and in the same perfection, if possible. This is opposed by a sense of learning *as* experience (cf. Meyer-Drawe 2003). Experiences can neither be planned nor controlled with a curriculum. They are occurrences (Waldenfels 2004a, p. 66) that happen along the way, in which the unforeseen and out of line becomes an eye-opening experience or—a prerequisite for "learning as relearning" (Meyer-Drawe 1986)—thwarting previous assumptions about the world and things. Only when learning is released from the corset of the canonized (and appropriately evaluated) result can learners be perceived in their heterogeneity and not as deficient concerning a predetermined target goal, but in their being different, being unique, and being the way they are. It is not the learners who have to adapt to a concept of normality in learning, but rather the ideas of normality have to widen so that the given reality can be expressed and shaped (cf. Peterlini H.K. 2018).

Perceiving learners and learning itself as different requires an open attitude of receptivity, which does not start from expected achievements and measures learning based on it, but instead can be surprised by learning as kairotic event in the sense of an aha-moment. In search of a concept of learning that may grasp this event "beyond" the given, guided, and expected, Schratz (2009) suggests a "learning-side" perspective (cf. also Schratz et al. 2012, pp. 21–30). A "learning-side" view includes teachers and learners and "is consistently based on uniqueness, i.e., the learning experiences of the students" (Schratz 2012, p. 18). This requires and enables the broadening of the understanding of normality for a perception of difference. Meyer-Drawe added:

> *"From the perspective of finished value orders, the contribution of a child's worldview is made to disappear only as a deficit. In a phenomenological analysis of the upbringing and teaching process, this is not addressed from the end goal, i.e., the relevance of the postulated goals, but from action as an implementation of meaning. From this point of view, pedagogical action appears more like a conflict process than an investment system."* (Meyer-Drawe 1987, p. 71)

This understanding of pedagogical action "more as a conflict process than an investment system" (ibid) breaks the established hierarchy between teachers and learners. In a conflictual process, both sides can no longer be sure of their position. It requires an engagement with the other. How one or the other person ever presents him or herself is not predetermined. Being different constitutively eludes predictability, which may explain the need for security and the desire for homogeneity of school classes described at the beginning of this chapter. One of the other is, by definition, *different* than expected. On the *teaching side*, this means giving up secure positions in the spirit of an open exchange on the level of relationship and mutual learning, replacing the pedagogical hierarchy. Educators may be encouraged by enriching their understanding from and with pupils that Meyer-Drawe described as "teachability of the teacher by the learner" (ibid). The heterogeneous school class is then a learning field for receptivity *and* self-activity. It does not require homogenization but has its value in the diversity of the learners.

Dialogue with Adorno. *How to Deal with Right-Wing Populism, Racism and Institutional Cruelization*

> *"Germany is responding to Turkey's threats to open borders for refugees with a tough deportation plan. [...] Austria –currently number 1 in the E. U. for deportations anyway –is following suit."* (Oe24 2016)

> *"And when do they shovel out more than 1.5 million, when 2 thousand deport each year and more than 300 thousand come in? I'll crack those riddles right away. Never!"* (Ibid, blog comment)

> *"I also dare to say that all conspicuous people are Muslims, that must be allowed,' explains the State Police Director Andreas Pilsl."* (Kurier 2017a)

At the beginning of this chapter, there is a feeling of pedagogical powerlessness. What can we do in the face of the stubbornness with which racism and violence against vulnerable people assert themselves and spread despite so many pedagogical concepts and efforts in the last decades? So this text could cue the violent and deadly mistreatment of African-American citizens by the police in the USA in early summer 2020 as the low bright light in the dark continuum of racism. From a European perspective, it would be (too?) easy to point the finger at the brutalization elsewhere. Thus, the focus of this chapter targets much less spectacular incidents and attitudes in supposedly civilized Europe, the so-called cradle of culture, democracy, social solidarity.

This image has become increasingly cracked, at least since 2015. We are familiar with these events, even if they have become more ephemeral again: the increasing number of migratory movements towards Europe, the images of the 4000 children cut off in Aleppo at Christmas 2016 and exposed to the risk of death, the drowning people in the Mediterranean and the rejection of lifeboats in Italian ports, the cynical strategy of most European states to abandon the refugee issue to the exposed southern countries and even to pay repressive regimes for barracking refugees in humiliating conditions. Ultimately, the multitude of harrowing news and comments lacks understanding and co-experience. Even refugees from Afghanistan after the dramatic war development in 2021 found a cold refusal at Austria's boundary.

The hostile stance of European refugee policy relies on a vulgarized public discourse that conversely is nurtured and legitimized by a policy of harshness. In the increasingly digital media world, consternation can be expressed with *Likes* and *Dislikes* but remains lonely because it is only shared virtually. After all, consternation is challenging to convey from afar, in which everyone can post hate messages without feeling the hurt on the other side.

There is a dividing line between supposed facts and evaporated emotions, a gap between concrete experience and communicative reproduction. It also separates pedagogical thinking from concrete options for action. Educational science faces pressing expectations to provide strategies, tools, methods, bags of tricks, curricula about education for and mediation of peace and against the social dislocation. Despite comprehending these demands, it is necessary and valuable to clarify the objectives and possibilities for pedagogical action beforehand. What requirements do xenophobia and racism take for the *art of education* in the perspective of peace research and peacebuilding? What can this science do, how much does it promise, and how little does it keep? What should pedagogy and even education for peace do in the first place if its efforts are counteracted by the sheer cold-bloodedness from politics and media, legitimizing institutional racism at a high level? What does it mean for the approach of *learning based on the model* popular in peace education (cf. Jäger 2016, p. 25) when the model from which people should learn is a politic of coldness? In Austria, i.e., a minister prides himself on the refugee crisis that Austria is a "European Champion in Deportation" (*Kurier* 2017b), and he will reduce the upper limit for asylum seekers again by half, so to speak, to have set a new record of refusing to help? (ibid) What kind of de-solidarization and brutalization of a society is based on the "model" of a political leadership class if another minister "acts on deterrence" by "rigorously intercepting refugees in the Mediterranean, then immediately sending them back or interning them on islands like Lesbos" (*APA* 2016). The news that an "apocalypse" is threatening Aleppo (*Krone* 2016), which already removes real pain through the metaphor, promptly followed the subtitle with the cynical and panic-raising question of whether "another million refugees" would come with it. In the blog, the first entry read: "... no problem, but please no longer to Austria!!" (ibid) Against all resignations that may go hand in hand, the question is asked: What can pedagogy do despite this if it sheds the illusion and removes the allures of the feasibility of education and learning, and admits its weakness, its impotence, to turn this into its actual strength?

The Everlasting Auschwitz as Challenge for Education

This chapter tries to dialogue with Adorno's essay "Education after Auschwitz" (cf. Peterlini H.K. 2017a). Adorno wrote this fundamental article in 1966, about 25 years after the establishment of the Auschwitz concentration camp, nearly 20 years after the end of the Second World War and the Holocaust, seeking to answer the question of how education can prevent this in the future (Adorno 1966/2010). As a tribute to Adorno's epochal text, this chapter does not want to compare current wars, extermination strategies, and racism with historical Auschwitz. Still, it does refer to the everlasting Auschwitz that Adorno also had in mind when he wrote:

> "One speaks of the threat of a relapse into barbarism. But it is not a threat – Auschwitz was this relapse, and barbarism continues as long as the fundamental conditions that favored that relapse continue largely unchanged. That is the whole horror. The societal pressure still bears down, although the danger remains invisible nowadays. It drives people toward the unspeakable, which culminated on a world-historical scale in Auschwitz." (Ibid, p. 2).

The level at which pedagogical science, then and now, is overwhelmingly challenged lies in the split between information and consciousness. It is the crucial point for historical Auschwitz and shapes the current way of dealing with destruction and contempt for human beings, dehumanization, and cynicism. In "Dialectic of Modernity" (Beilharz 2000), Zygmunt Bauman tried to explain the Holocaust, based on Horkheimer's and Adorno's "Dialectic of the Enlightenment" (1944/2002), as a consequence of the enlightenment. The rationalizing and disciplining structures of order, work, and economization contributed to the division between reality and consciousness in modern societies. In interaction with the complex causes of the Holocaust, this division of consciousness represents the enabling of the horror. How else should we explain the detailed planning, precise organization, and consistent execution of mass murder by principally emotionally capable humans? What made the genocide possible can make it viable again at any time. It is not enough to know the bestiality of war elsewhere, not even the cruelty to the neighbor next door if this is split off from conscious tracking and experiencing. Knowledge, information, and instruction are essential but not enough. The experience is needed that makes sensing possible and sustainable.

Here pedagogy faces the limit that makes one perplexed and yet represents its fundamental task: How can people be moved, challenged, accompanied to feel themselves so that the suffering of their fellow human beings does not bounce off of them so indifferently? Is this even possible? Why is empathy moving so many people and, in turn, so little or not so pronounced many others? What allows the last one to entrench behind stereotypes in front of manifest misery and perceive others only as stereotypes? What attitude can education adopt towards those who unconsciously work in barbarism in the words of Adorno? They would be the primary target group for education and training, but at the same time, they seem unreachable because they imply pedagogical failure and are a source of impotence. Beyond this level of the individuals, pedagogy has also to consider the implications of social and political orders. Why do superordinate systems like states, which ultimately emerge from human practice, act in warlike, exploitive, and destructive ways?

The culminating increase or even new social admissibility of xenophobia in everyday media and political discourses fundamentally challenges peace science in all its disciplines, orientations, and ramifications. Working for peace requires transdisciplinary approaches as a transgression of epistemic dividing lines and decolonizing scientific dominance discourses. However, this does not free individual sciences, including pedagogy, from responsibility in their field and within their paradigms. From the perspective of cross-discipline peace and conflict research, a complex analysis of social, political, socioeconomic, cultural conditions, structures, and dynamics is required. For pedagogy, like any other science, to get involved, it has to clarify its tasks and roles, its limits, and possibilities.

Everyday Racism on Train: Have We Learned Something?

In this chapter, answering this question attempts to determine how people and social groups can learn sustainably, deal better, more peacefully, and more fairly with one another. To do this, they have to learn to clear or even overcome enemy images, put down patterns of defense and rejection, and structures of subjugation and exploitation, to open up to convivial living together. Following the Balkan wars after 1990, Alexander Langer adopted these thoughts in a policy aimed at regional and national reconciliation interpretation of the conviviality concept by Ivan Illich (1973) (cf. Langer 2015, Peterlini H.K. 2015).

A dilemma becomes apparent: There are well-proven models for the inter-and transgenerational transmission of images of hostility, patterns of hatred, dynamics of subordination, classification, and superiority through myths, traditions, rites (cf. Assmann J. 2011), which in their adherence to closures plausibly explain their durability. In contrast, the sustainability of learning that opens up or at least stretches such patterns for relearning and further learning seems to be based on a theoretically and praxeologically fragile basis. Such historical moments and developments of transcending are evident in the historical review and present-day findings. Passing them as proven and preserving behaviors of the good life and peaceful coexistence is exposed to permanent setbacks or relapses, as Adorno would call them. According to Aleida Assmann (2015), making more fluid the sedimented firmness of cultural memory may help to deconstruct traditional patterns and roles but ultimately leaves open *how* they are then newly constructed, designed more flexibly, or hardened again.

> *Conversation on a train en route from Austrian East Tyrol to the Italian South Tyrol: an elderly couple from East Tyrol and an older woman from South Tyrol talk casually of train conversations between strangers, brought together by the vacant seat. The woman from South Tyrol says that the young people no longer want children in her circle of acquaintances or, at most, one. Recently one of these young women said that she would rather have a dog, "I told her, aha, then the dog will probably pay you the pension – she was quiet there, she didn't say anything more." Then, almost without transition, she says that in her village, the locals are being sued out of the apartments because they can no longer pay the rent on time. "And then they put the foreigners in because the provincial government pays for them, so the money comes on time – this is how it works here in South Tyrol." The man from East Tyrol interrupts her, "well, that's not only the case in South Tyrol, but it's also exactly the same with us. The foreigners get everything". "Really," the woman from South Tyrol marvels, "my God, then we will soon only have such people here, well, well." The conversation falls silent, then there is talk about illnesses and that the doctors only look at the computer when they treat you. (Note, December 12, 2016).*

> *On the Kufstein-Brenner route, from Austrian North Tyrol to Italian South Tyrol, two young couples, one from Germany and one from Italy, came side-by-side. The two women sit in the middle, and the men talk to each other across from them. The young Italian man talks about the mafia in the area where he comes from, the German, who speaks Italian somewhat, interrupts him: "La Mafia anche da noi, si chiama Merkel" – the Mafia is here with us, she is called Merkel, her foreign policy is a crime against Germany, "sempre più nuove persone che non fanno bene" (more and more new people who are not suitable for us). The two young men get into a more heated conversation. The women were no longer safe, and the foreigners earned by drug trafficking. "The mafia has something good there," says the young Italian, "it has the best method. If the negretti [little negroes] trade drugs, then … bang, and they come in the cement for the shell. Yes, why use cement when you have the negretti." The German laughs, the Italian agrees. (Note, November 10, 2016).*

Two generations, one growing up with the tragedies caused by National Socialism, the other beneficiary of post-war European peace, in which war is only known by media. The racism of the elderly sounds vague. Just as you talk about neighbors and youth nowadays who prefer to have a dog over a child, you also speak about foreigners who get everything, while the "own people" go away empty-handed. The racism of the younger generation is more pronounced, it complains about the crimes of the mafia and places the German Chancellor at the level of organized crime because of her refugee policy, and it praises the mafia murders for executing unwanted henchmen in its own core business of drug trafficking. Both generations speak openly on the train what years ago would have been whispered ashamedly or only at a group gathering's table, at a late hour, and with a tongue loosened from alcohol. How far is it from such a speech, which is omnipresent in social media, to the appropriate action? And what is the difference between "negretti" and the working term "Nafri" used by the German police for North African young men? The splitting patterns between the good "We" and the evil "Others" are evident. There is the general suspicion of foreigners that they come to publicly sponsored apartments at the expense of the locals, while in the same country, namely Italy, cheating the state by tax evasion is more than in use. And there is the mobilization against sexual harassment by a single group, the Nafris, while sexual assault and sexist behavior is taboo topic when it takes place in local groups and families. Has there ever been a lesson learned in recognizing and overcoming racism? Or have we fallen back into that cruelly real fiction of everyday fascism? The cabaret artist Qualtinger had condensed it in his fictional figure of *Herr Karl*, which tells in a chatty tone that it was only meant as fun when Jews had to clean the cobblestones with their toothbrushes. (Merz and Qualtinger 1996).

Pedagogy Cannot Replace Politics, Yes, but then What Can?

For Adorno, the "premier demand upon all education is that Auschwitz should not happen again" (Adorno 1966/2010, p. 2). As a key text of critical-emancipatory educational science, his essay expresses the often overwhelming hope for academic strategies and educational measures, as well as the fear of their failure. Education has to know its limits if it does not want to be fooled by the—constantly flashing—successes or the—always threatening—losses. Niklas Luhmann and Karl Eberhard Schorr (1982) described this in the language of systems theory as a "technology deficit." Phenomenology expresses it in a more friendly but no less sobering way for every pedagogical feasibility belief: "When teaching, we put the success of our actions in a foreign hand." (Waldenfels 2009, p. 32) For this insight that the linear implementation of pedagogical measures is ultimately an illusion, Siegfried Bernfeld (1925/1973) had long before named three limits of education in his "Sisyphus": first and foremost, that social or societal limit that deprives educational interventions of any monocausal certainty of results.

In his considerations of how education can prevent Auschwitz in the future, Adorno also refers to this limit when he admits "that the recurrence or nonrecurrence of fascism in its decisive aspect is not a question of psychology, but of society" (Adorno 1966/2010, p. 3). He speaks of the psychological, with which Adorno meant the conversion of destructive unconsciousness into reflected consciousness in the psychoanalytic sense "only because the other, more essential aspects lie so far out of reach of the influence of education, if not the intervention of individuals altogether" (ibid). In this way, pedagogy cannot replace politics, as Franz Hamburger (2010) postulated concerning inequalities in the migrant society. However, conversely, pedagogy must also become aware of its political dimension and re-politicize by recognizing social, economic, patriarchal, structural patterns of violence in society and considering the pedagogical approach. The fact that pedagogy cannot replace politics does not mean that pedagogical action can leave politics in charge alone.

How then can we have an educational effect without waiting for the utopia of social change or focusing exclusively on the field of political debate? Compared to political debates and practices, the voice of pedagogy is usually weak or manipulatively used in the service of more powerful discourses. In this sense, Adorno's reasoning has remained topical: Those who cannot change the social structures that require individuals to suppress their desires—"a life of human dignity" (Adorno 1966/2010, p. 7)—and mutilate their consciousness (ibid, p. 4), have to start where Bernfeld has identified a further boundary of education: namely with the child itself, more openly formulated with the object-subject of education, which however is resistant to technical interventions that believe in causality due to its different history, constitution, and plasticity.

Here we are in front of a pedagogical dilemma. On the one hand, the subject itself is socialized and subdued to the social order (which would have to be changed). On the other hand, it is "yes always the individual" in educational thinking (Parin 1999, p. 170) with whom the dispute should occur. According to such a claim, children, adolescents, and adults would have to be repaired to adapt to social and political requirements.

Pedagogical action stands here in a field of tension between subject and world, in which the educators themselves (in the broadest sense) are involved. The pedagogical impetus thus also comes up against that third limit, according to Bernfeld (1925/1973). It represents the educators themselves because of their own previous experience (the child they were themselves) and socialization conditions. No action is possible without self-referential intent if they do not exceed previous individual and socialization experiences in a reflected manner. Consequently, educators' encouragement to self-reflection and self-awareness would be the first step for educational actions if they should have any pedagogically meaningful effect.

The Problem of Wanting to Educate for the Good

But what does it mean to be pedagogically *meaningful*? Who decides what is meaningful? Education to renounce violence seems like a logical answer. However, it's not so easy. As a normative approach, it hides the conditions in which violence arises, in which violence can be an outbreak of impotence, a response to manifest or implicit violence (see, among others, Heitmeyer 1994) or, as in many cases, military interventions as a prerequisite for peacebuilding (see Lakitsch and Steiner 2015). From a skeptical pedagogical perspective, Heinrich Kupffer, as explained in the chapters before, denies that education can be carried out for a particular political model at all. Though education for peace is an even more general value than education for democracy, and therefore a normative setting can be plausibly represented, it could hide the pedagogically problematic design of the new human, onto whom objects or subjects of education are trained. For example, Montessori's reform pedagogy, often cited as early approaches to peace education (cf. Reitmair-Juárez 2016, p. 184; Pistolato 2016: 164), drafts unmistakably racial-biological and eugenic traits (cf. Hofer 2010). These are rarely or only shamefully discussed in peace education and educational science in general. The question of how pedagogy can contribute to promoting peace can be answered more conclusively with John Dewey (1916/2009) and with transformative approaches according to Paulo Freire (cf. 2007). However, both approaches, as different as they are, attribute learning to experience, which—with Dewey through non-simulated, but rather a concrete action and real problem solving, with Freire through dealing with one's own life situation—is challenged, stimulated, provoked, but cannot be controlled. Experiences happen; they can't be ordered pedagogically. And even if we initiate or incite experiences in performative settings, we have no control over how people live the provoked experience and what they learn from it.

Education science with the normative claim to educate *for the good* is facing the double fundamental problem of a discipline, which on the one hand should contain manipulative access to the subject and on the other hand may initiate, reflect and observe processes, but has no certainty of results. No matter how good, every educational intention has to struggle with the dilemma "between results orientation and the openness of results" (Frieters-Reermann 2016, p. 63). Peace education can provide impulses, impart knowledge, circumscribe curricula, try out interventions, practice attitudes and explore learning methods. However, as in any educational or teaching process, the actors must be aware that they cannot steer the result. This insight means to say goodbye to the idea of didactic feasibility since all learning *as* experience (Meyer-Drawe 2003) may or may not take place in its unique way only in the subject's confrontation with the world and people, their limiting conditions, and empowering potentials. Concerning studies on environmental education (Kuckartz 1998), Fritz Reheis also notes that there is hardly any direct influence on behavior from imparting environmental knowledge for peace education (Reheis 2016, p. 34).

School can't do it Alone, but it can't be Left Alone Either

The openness of educational results is not due to the techniques of teaching. Delegating integration, peace education, peacebuilding, and learning peace to school didactics and curricula alone is, on the one hand, equivalent to overwhelming schools and, on the other hand, deflects responsibility for society as a whole. Ivan Illich's provocation that society needs *de-schooling*, as schools reinforce the inequality structures at the expense of the already disadvantaged (Illich 2002), has not lost its stimulating sting. Of course, there are ambivalences to take into account. The school is a critical institution for integration and social inclusion since it also selects but has a place for everyone anyway up to a certain age and, despite selection mechanisms, still visibly represents heterogeneity. Through the presence of others, schools contribute a little to their normalization (in the above-discussed meaning). But there are limits: Trusting that curricula enable learning about peace is naive as long as the school itself reflects social division and devaluation patterns and is incorporated into social, political, and state structures of division and disadvantage. No lesson can make up for the inequality, discrimination, social ostracism, and degradation, forms of overt and subtle violence that are overpowering outside of school, have an effect on school and at the same time are produced and reproduced there (Peterlini H.K. 2016a, p. 52).

Accordingly, Adorno's proposals for the upbringing and educational measures against the return of fascism go beyond school and nevertheless assign responsibility. The suggestions are not without eager helplessness:

> *"I can envision a series of possibilities. One would be — I am improvising — that television programs be planned with consideration of the nerve centers of this particular state of consciousness. Then I could imagine that something like mobile educational groups and convoys of volunteers could be formed, who would drive into the countryside and in discussions, courses, and supplementary instruction attempt to fill the most menacing gaps. I am not ignoring the fact that such people would make themselves liked only with great difficulty. But then a small circle of followers would form around them, and from there the educational program could perhaps spread further."* (Adorno 1966/2010, p. 4).

And finally: "All political instruction finally should be centered upon the idea that Auschwitz should never happen again." (Ibid, p. 8) Looking back on decades of historical education about National Socialism, democracy, and peace-promoting initiatives on many levels, in schools and the media, society, and associations, the—at the time—subdued hope should now result in resignation. On the one hand, while the individual is banned from aggression, everything that was also a sign of return for Adorno continues to exist: the militarization and unbroken threat to the world through armaments and nuclear weapons (ibid, p. 2), the new, weaponized language of politics, which seems to have forgotten Auschwitz and which prepares the destruction mentally and verbally. Under formulas such as Frontex and border management, the governments' reactions to the refugee movements also serve as models for interpersonal boundaries and the social coldness mentioned by Adorno (ibid, p. 1 f.). On the other hand, everything that has been a breeding ground for barbarism for Adorno flourishes, partly in a new design and spiced up in a performative way:

> "One must fight against the type of folkways [Volkssitten], initiation rites of all shapes, that inflict physical pain—often unbearable pain – upon a person as the price that must be paid in order to consider oneself a member, one of the collective. The evil of customs such as the Rauhnächte and the Haberfeldtreiben[1] and whatever else such long-rooted practices might be called is a direct anticipation of National Socialist acts of violence. It is no coincidence that the Nazis glorified and cultivated such monstrosities in the name of 'customs.' Science here has one of its most relevant tasks. It could vigorously redirect the tendencies of folk-studies [Volkskunde] that were enthusiastically appropriated by the Nazis in order to prevent the survival, at once brutal and ghostly, of these folk-pleasures." (Ibid, p. 5).

Expelling Aggression not as a Solution, but as a Problem

Adornos criticism against rough traditions should be discussed further: In customs, psychoanalytical understanding shows the processing of instinctual needs, which are probably only perverted due to repressive, socially enforced instinct

---

[1] *Rauhnächte* and *Haberfeldtreiben* are customs beyond the German-speaking countries. The Rauhnächte usually extends over the whole Christmas Eve period. There are hauntings by ghosts, diseases, and wild demons, depending on the local arrangement. In many cases, however, the custom lives on in the donation of incense in all living rooms, cellars, and even stables to ask for blessings. The Haberfeldtreiben was often directed against people accused of bad manners and degenerated lightly into a kind of tribunal without rights of defense for the accused.

suppression and acted out as violent behavior. The election success of parties that brutally claim frowned upon social behavior and are thus successful despite (or perhaps because of) indignation at an established discourselevel at least point to such a connection. The election of Donald Trump as US President in 2016 and his unholding popularity even after the lost election in 2019 showed that the voters did not punish him for sexist, racist, vulgar, crude gestures, but probably even rewarded by certain strata of the electorate. The rude manner appealed to their instincts and social frustration at being despised and oppressed by the so-called educated class *(Bildungsbürger)*. Right-wing populism attacks precisely those values that required arduous intellectual debate—women's equality and rights, the racist taboo, the inclusion of the disabled, homosexual rights, and the liberal arts. The same can be seen in Italy in the right-wing populist Matteo Salvini and his *Lega*. They scored points in the 2018 election campaign and after his (short) entry into the Italian government with racist and inhuman statements and behavior. Despite embarrassing political missteps that catapulted him out of the government in 2019, he lost very little of his popularity.

Since the enlightenment has forced individual and social processes to rationalize, pedagogy has to question its relationship to instinct, aggression, emotionality, and even irrationality. Discipline and socialization in the sense of politically correct and socially acceptable behavior occur at the price of suppressing primal needs, which react gratefully to valves as soon as they are available. Likewise, the high level of violence in the everyday entertainment program is an indication that violence in the modern age may have been taboo and largely monopolized in state organs, but instead suppressed and split off than integrated and sublimated. This is only possible if people find design options in their lives and the living environment that values their vitality and creativity and does not suppress them.

The Latin origin of the word aggression—agredere—bears traces of consequential meanings, namely self-asserting, phallic, covetous participation. A society that collectively defies aggression but exclusively reserves it—in the above sense— for privileged groups (men more than women, economically favored people more than the disadvantaged, influential people more than poor) should not be surprised if the repressed comes back grim-like.

As also discussed by Adorno elsewhere (1966/2010, p. 5, 8), Pedagogy is thrown back on the usually quickly repressed question of how to handle libidinal, emotional, and physical needs. They should not be cast aside and suppressed but instead, be recognized for an actively shaped life in participation and empowerment. Physical education and movement pedagogy are mostly only alibi-offers by educational institutions that focus on the head and either ignore or discipline the body.

The assumption that xenophobia and racism can be fought with rational arguments alone fails daily. No doubt, it is necessary to try this too to counter the fears (cf. Bauman 2016) and prejudices about migration with facts. However, people persist in xenophobic positions who are resistant to facts and cling to conspiracy theories that are not accessible to rationality. Educational efforts beyond cognitive enlightenment are needed.

The same applies to the "anger against civilization" (ibid, p. 95) in a society administered close to claustrophobia without being a fair society. Adapting through self-discipline, rationalization, and repression of instincts is a duty for the economized subject. The reward for this effort is reserved for the privileged. Then as now, anger is not directed towards those responsible for social coldness, social tightness, and economic inequality, but is acted out on the weaker, with "violent and irrational" rebellion (ibid). "Post-truth" as the word of the year 2016 (Oxford Dictionaries 2016) has become almost a seemingly new phenomenon in current media and some scientific discourses, but the underlying human argumentation and behavioral structure is anything but new (cf. Setsche 2016). There is also a risk of it becoming a misleading phrase because "post-truth" indirectly asserts an essentialist claim to validity in the knowledge of reality. The blinded would only have to be convinced again by facts so that everything comes back into balance. Such assurance of truth is difficult to maintain, knowing the limitations of knowledge. The task and possibility of science, according to Habermas, is not to uncover the knowledge that is free of interest and unconditional but to see through the interests and conditions that represent the justification of knowledge (Habermas 1971).

The Disintegration of Discourse in the Digital Network—or the Inactivation of Speaking

The conditions under which individual learning and social negotiation processes are currently taking place have become more complex. Adorno's desperate idea that extensive education and information campaigns must penetrate the population would dissolve as an illusion in the digital world due to the often non-reflective, emotionally moved, and factually impossible to capture the spread of anger and fear. What Adorno attests to technology for his time and also prophetically diagnosed ahead of time for digital technology that he did not know is truer: "On the other hand, there is something exaggerated, irrational, pathogenic in the present-day relationship to technology" (Adorno 1966/2010, p. 7), which exaggerates the means of fetish and deprives it of its useful purpose. Here Illich could call for the utopia to reverse the enslavement character of the tools on a depth level: "The crisis can be solved only if we learn to invert the present deep structure of tools." (Illich 1973, p. 10) It is puzzling how this transformation of the relationship between man and "machine" can succeed.

One possibility is that people can relate to the machines, which, according to Illich, are not only technology but also our social institutions and structures. Where people establish relationships to things, devices, facilities, and institutions, they can transform the one-sided dependencies into exchange processes. Antonio Gramsci postulated this with the statement that "all men are intellectual, but not all men have in society the function of intellectuals" (Gramsci 1928–1937/1971, p. 9). To give people back their intellectual role, workers should not only stand at the assembly line but also have the opportunity to understand the background, social and economic conditions, and the cultural values of their work.

Only in this way could the tools be taken into the service of a good life. The Convivialist Manifesto, edited in 2014, claims for such a development in "the quality of our social relationships and of our relationship to nature" (Adloff 2014, p. 6). ood coexistence also requires the conditions that enable people to perceive each other as human beings, exchange ideas, and feel their fears, hardships, pains, expectations, hopes, and gifts. It's like a substrate of technology criticism both from Adorno and Illich. With Adorno, consciousness is covered by a "veil of technology" (Adorno 1966/2010, p. 7) that cuts people off from the feeling that— if they could deliberately deal with—would otherwise have to stir and get excited.

A picture of the social networks: The cold of the first weeks of January 2017, which was life-threatening and fatal for many refugees and homeless people, was countered by the coldness and malice of comments on the (alleged) failure of the welcoming culture and goodwill. Just a few quotes from a Facebook debate on a journalistic report titled "Refugees: Without protection against cold and indifference." (*Die Zeit* 2017):

> *"Go home where it is warm or to Saudi Arabia etc., but you have to work there. You get no financial help from the rich Saudis. You will be driven back even by force; here you have to eat pork or starve to death." (With name).*

> *"We are not the catch-all for all misery and free-rider drivers in the world ... Everyone is already working for 30 refugees a person." (With name).*

> *"I only see young men. They should stay and fight in their country and not rape the women here. Can't hear all of this anymore ... Yes, then I'm a Nazi, I don't give a shit !!!!" (With name).*

> *"If you don't want to get on the ship, it's up to you whether you freeze, I think." (With name).*

The rampage of racism and xenophobia in digital social media is no accident. Here the debate is removed from all social control. While the web-redaction of *Die Zeit* deleted a large number of comments on the quoted post (ibid), these continued to be uncensored on Facebook. Disinhibition cannot be reduced to digital, everyday fascism—as could be documented daily—it also expresses itself in unusual rawness among the general public. With the Internet, however, a reinforcing medium is available that has radically democratized and delimited public speaking in a kind of private–public: not only the educated and privileged, to whom the academic language, newspaper columns, microphone, and camera are available, but almost everyone (with internet access) can directly, without control, without censorship, without responsibility and thus primarily safely and shamelessly express their opinion to a not so small public, which nevertheless conveys the feeling of communication "among friends." At the same time, this speaking takes place in a room without bodily resonance, without emotional repercussions for those who speak, and thus beyond empathetic relationship and democratic discourse. A new dimension of public space has emerged on the net, which—as Byung-Chul Han argues in dialogue with Habermas's communication theory— defies the rules of public discourse (cf. Han 2013, p. 16). Speaking to Habermas: "For the present, there are no functional equivalents, in this virtual space, for the structures of publicity which reassemble the decentralized messages, sift them, and synthesize them in an edited form." (Habermas 2009, p. 158).

The Internet escapes pedagogical or educational controls. It obeys the logic of swarm intelligence (cf. Surowiecki 2004), for which it cannot be said with certainty how it plays out; whether it leads to *swarm democracy* (Han 2013, p. 11) or to the "decay of public space" and thus democracy (ibid). It isn't easy to decide whether the de-hierarchization of communication only reveals what would otherwise be stealthily thought until it violently paves the way at some point. Or whether the medium—discussed by McLuhan (1962) on book printing—is itself the message that creates the reality it represents. Han (2013, p. 41) cannot decide a priori whether this reality will be "a utopia or a dystopia, a dream or a nightmare."

Undoubtedly, the *ideal of hardness* criticized by Adorno (Adorno 1966/2010, p. 5 f.) finds vast space in digital communication. Distance and virtuality desensitize the exponentially increased exchange, impacting public discourses outside the web. Although the network is beyond control, the speech outside does not escape from the disinhibitions safely tested. This probably changes the discourse per se and deprives it of critical quality of exchange, namely that of comprehensibility in the sense of empathy, so that we can comprehend and feel what the spoken or written word causes in the recipient.

In the digital public, the effect of one's word is not noticeable, so that anger and fears can be discharged but ultimately lead to nothing. The disposal gets stuck halfway. There is no discursive exchange, just an exchange of blows. Analyzing blogs, we can understand how the tone usually intensifies because the response is never comprehensible, and neither bodily nor sensually can't be. However, some are insulted and disparaged. Their dismay, blushing face, tears, anger are invisible to those who attack and write back. Something similar concerns, of course, the traditional media, in which political discourses are almost as enclosed as under glass. Statements concerning humanity questions are often spoken in microphones and cameras without sensitivity to what these words do to those who hear them and whose faces.

The Appreciation of Fear in Response to its Postponement

The disembodiment of political and media speech contributes to the hardening of the discourses and the thereby justified actions. The coming together of masochism as discussed by Adorno (having to put up with the hardness of others, the hardness of life, the need to adapt to economic constraints and structural subordination) with sadistic acting out could also be a description of the present:

> "Being hard, the vaunted quality education should inculcate, means absolute indifference toward pain as such. In this, the distinction between one's own pain and that of another is not so stringently maintained. Whoever is hard with himself earns the right to be hard with others as well and avenges himself for the pain whose manifestations he was not allowed to show and had to repress." (Adorno 1966/2010, p. 5).

What prompted Adorno to plead against an upbringing attitude that "sets a premium on pain and on the ability to endure pain" can also be related to more subtle constraints. We find these in educational structures that demand and promote adaptation to (social, economic, cultural, patriarchal, hierarchical) conditions. They punish at the same time feelings of reluctance, disobedience, stubbornness, frustration against the associated offenses with a degrading measurement of the subject, be it through grades in school or through social gradation and exclusion in social and economic reality. The processes by which the subject finally internalizes external devaluation as self-devaluation have been extensively researched and described, from Freudian psychoanalysis to Foucault's techniques of the self and Bourdieu's theory of habitus.

Adornos conclusion for parenting and educational understandings was and are obvious:

"Education must take seriously an idea in no wise unfamiliar to philosophy: that anxiety must not be repressed. When anxiety is not repressed, when one permits oneself to have, in fact, all the anxiety that this reality warrants, then precisely by doing that, much of the destructive effect of unconscious and displaced anxiety will probably disappear." (Ibid).

The most apologetic indication of right-wing populism on fear of strangers is debatable and not very profound. At the same time, a pedagogical perspective should consider that worries do not need an objective ground to be excruciating and tempting to act out. Fears often are shifted from quite concrete causes to phantasmagorical delusions. Then they hit those who may not be responsible for them but serve as projection screens or projection figures for disposal of fear.

The dynamics of repression do not allow everyone to relate their fear to concrete life conditions, like social decline, uncertain economic future, professional defeats, economic destabilization in a competitive society, physical frailty, and death due to the constitutive vulnerability of human existence. These fears are dumped on those who represent all this and at the same time allow a distance— the even weaker, the disadvantaged, the impoverished, the ragged, the failed, as well as the refugees in their otherness.

Ultimately, these are "messengers of the repressed" (Berghold 2005, p. 111) who have been punished for awakening the memory of the repressed fears. In an interview, Zygmunt Bauman (2014/2017) defined the strategy of Facebook founder Mark Zuckerberg as a business based on the fear of loneliness, which expresses the existential concerns of being exposed to humans. This enables "a lot of communication, but no dialogue" (ibid), the fear finds valves, but no processing of reality that can be experienced and no interpersonal comfort.

If Adorno, as a condensed statement on upbringing after Auschwitz, believes that "the only education that has any sense at all is an education toward critical self-reflection" (Adorno 1966/2010, p. 1), an intermediate step would have to be inserted. To enable critical self-reflection and promote it through pedagogical and educational strategies, people must learn to feel themselves and connect with their fear, pain, and insult. Only then can the play with fear, with the removal of fear through disinhibition in the projection of others, maybe not be thwarted, but be avoided here and there in favor of examining what hinders people from a good life in their real condition. In line with Gianni Vattimo's (2013), weak thinking requires "weak pedagogy" discussed above. Such an attitude does not trust secure worldviews but knows about the precariousness of human existence, a pedagogy that is not based on a priori principles but tries to understand (not justify) human narratives in their captivity and limitations. A "weak pedagogy" knows about its ethical standards because it has reflected on them but takes them back in pedagogical action. In this sense, it is not weak. Still, it makes itself soft, not from a strategic calculation to pretend communication, but to flatten the asymmetrical gradient in the pedagogical relationship. From the perspective of education's subject, the proclaimed values often come from a hierarchical stance. This moral posture is a strong temptation but is pedagogically sterile when dealing with undesirable phenomena, such as racism and nationalized or religiously fanatical identity.

In schools, it can often be observed how teachers flinch when students stand up and take up racist positions. The logical contradiction weakens the teacher to the point of helplessness and strengthens the defiant rebellion and adherence to the irrational (Peterlini H.K. 2011, p. 169 f.). If you want to talk to a child (and not down on them), you must sit on the floor. If you want to work with adolescents and adults who are right-wing populist or religiously fundamentalist, you can withdraw to a good-evil division of the world and turn away in disgust. Or you may deal with it, listening, asking, and not being shy of your answers. This discussion cannot take place virtually, which would not be possible due to the distance, but requires a concrete encounter, if possible, in protected and designed rooms. Because the public debate is also unsuitable for this, it takes place at the level of systems and strategic communication, which has the gain and victory over the other person in mind, *not the understanding of the Other* (Peterlini H.K. 2012). At the political level, the debate has to be faced. It consists of a civil society challenge to dominant discourses at the systemic level. In a pedagogical setting, however, the dispute does not downscale from above that wrong thinking manifests itself there even if it is won. It might help overcome impotence, but it would be equivalent to withdrawing from the field where pedagogy is genuinely required. In this field, educators have to face everything they do not want, what runs counter to them, what they try to prevent, and what shakes and challenges them in their basic understanding.

Deconstruction of Dichotomy as a Key Educational Task

Ultimately, such a pedagogical approach is about making dichotomous divisions in the perception of self and reality of life on an experience level accessible, perceptible, open to reflection and narrative, thereby weakening their rigidity. When divisions become perceivable, commonalities and differences can be thought of as a never definite plural and experienced as formable. Helpful on a theoretical level are the approaches of transculturality (Welsch 1999) and trans-difference (Allolio-Näcke et al. 2005). Wherever divisions pervade reality perception, it is seductive discharge fears, worries, anger onto the other half of reality that is different. There they remain locked up, inaccessible, and develop their threatening potential.

The senselessness that inhibits people from "striking outwards without reflecting on themselves" (Adorno 1966/2020, p. 2) requires, according to Adorno, not only an education for critical self-reflection (with the head) but also a more careful sense of the body (ibid, p. 5). To restore the mutilated consciousness, Adorno accepts a split at this point, which is itself the fundamental problem. The independent thinking and keeping apart of mind and body as a matrix of a dichotomy, which is on many other levels between "self" and "foreign" reproduced and always upgrades or downgrades one pole compared to its opposite pole. Considering that skin color and appearance impact who is how often controlled by the police, does this mean that physical details are viewed in isolation and separated from the subject. The other person is perceived as a "foreign body" (Nancy 2008, p. 5) and not in its entirety as an *animate body,* according to Waldenfels (2000, p. 14; cf. Waldenfels 2004b). The concrete person cannot split up according to distinctive features and is never fully grasped by nationally, culturally, sexually, socially constructed categories.

Accepting the other without negating the difference is an unconditional consequence of the insight into what can happen otherwise. The holocaust resulted from a conception of the others as entirely different, robbing them of any humanity and creating an enemy image, which could be exterminated without remorse (cf. Peterlini HK 2011, p. 49). According to Adorno, a "potential for enlightenment" would already lie in the effort that the subject recognizes its so-being in its historical and biographical development and does not mistakenly consider it as "nature," as something "unalterable given" (Adorno 1966/2020, p. 6). The naturalization of diversity inevitably biologizes and racializes the other to an immutable and irreconcilable Other.

The thinking away of the Other, the "expulsion of the Other" (Han 2017), is the mental step of the desire for its actual removal as it was practiced in the most extreme form as destruction in Auschwitz. It is also ultimately expressed indirectly in the criticism of bailouts in the Mediterranean, according to which it is better to let refugees die than to save them. In this way, Adornos's pedagogical postulate could be expanded from education to critical self-reflection to a perceptible and reflective exploration of one's way of being different from others. Only those who learn to perceive and endure their strangeness (towards themselves, towards the world, their supposedly own culture, towards their socio-economic structure) can accept strangeness and closeness by others. This asks for learning to think of belonging and difference no longer in absolute terms but in a broken and overlapping manner

On the one hand, it is about the understanding of the difference in its infinite diversity, as Arens and Mecheril (2010, p. 11) probably best expressed with the above-quoted axiom that "everyone is different different" (*anders* anders). Precisely because of its charm, we should reflect on this formula again. In principle, it questions constructs of equality. It does not perceive being different in isolated and special groups or specified deviation criteria in isolation (such as physically impaired, strong learners, weak learners, foreigners). The difference is potentiated and infinite. Thus indefiniteness is driven, which removes the categorization and discrimination foil from it. But there is a double-edged point of the hypothesis. Such a radically set difference could implicate the impossibility of community and group formation of peers who have to fight for recognition of their difference and rights.

It is crucial to ensure that neither difference nor equality is set absolutely but that the transitions and intermediate areas are appreciated. Just as "everyone is different" in a different way, we are also "different *alike*" and "*similarly* different." The *absolute Other* can be potentially destroyed because the subject is isolated and detached if all equality is left out. Absolutely set equality promises belonging but would seduce into suffocating symbiosis because it means that the delimitation that is indispensable for one's existence is lost. The concept of *equal different equal, and similar divers* move in a life-threatening and, at the same time, life-saving ambivalence. Difference and equality remain in limbo and have to be renegotiated again and again. This ambivalence neither creates complete division and loneliness, which exposes the supposedly autonomous subject to severe uncertainty, nor does it create unbroken belonging, which seems to convey the security, but would itself be suffocating in a symbiotic manner. Accepting this requires familiarizing yourself with your ambivalence of "differently *different*," "differently *the same*," and "*similarly* different." Gianni Vattimo's reinterpretation of Nietzsche's superman tells from people who are not superhumans but beyond humans (oltreuome). They accept weakened worldviews and learn to exist creatively, meaningfully, and responsibly in their existential insecurity (cf. Vattimo 2019).

The Lifeworld as a Space for Experience and Testing

Becoming *oltreuomo* requires learning by individuals and collectives. With the above-discussed dual-concept of system and lifeworld, Habermas (1984, 1987) juxtaposes two social dimensions in which, on the one hand, the individual can learn socially and act in a solution-oriented manner, and on the other hand, is subject to controls from dominant systems such as politics and economics. In the lifeworld, especially when it comes to everyday practical aspects (which Habermas always has to keep pervading), people are capable of communicative action, but at the same time are also impaired by the "colonization" of the lifeworld by the systems (economy, politics) (Habermas 1984, p. 46, 1987, pp. 196, 325, 356).

The model offers several starting points for reflecting pedagogy and education from a peace science perspective: Accordingly, the lifeworld can negotiate interests. This is not necessarily idyllic, it can also be conflict-ridden, but the interests of the individual must be socially dealt with in their interactions in a communicative manner (and not striving for strategic overreaction and, ultimately, the destruction of the Other). To perceive these processes of negotiating opens up a promising pedagogical field. How can education promote communicative action and make them pedagogically fruitful? In any case, the concept contrasts the idea of an anthropologically rooted tendency to destructive attitudes of humans, which makes hostility and war inevitable. People show that they can negotiate and solve problems together in the lifeworld.

At the same time, the effects of the systems compromise the conflict-solving competence of the lifeworld. Unlike the lifeworld, systems obey the laws of strategic communication with a dominant control medium. In the system of economics, this can be money, profit. In the system of politics, it is power. The dilemma of strategic communication is that it obeys either-or principles—either I make the profit or my competition, get the majority of the votes, or my opponent. This rule makes the communicative exchange more difficult and promotes supremacy. Only when the competitor is defeated, and the political opposition is beaten can one's survival be considered secure. In the final analysis, that is the logic of predatory capitalism in the economic system, which has discarded all social responsibility. It is the logic of a political majority that has no scruples about the use of force to achieve its own goals in the system of politics. The dilemma is that the systems affect the lifeworld and subject it to its steering media, colonizing it (Habermas 1987, p. 267). The model leed in a dichotomous pattern: On the one side, the solution-oriented living world, on the other the destruction-oriented system world.

Habermas suggests a way out of the dichotomy with his concept of discourse developed in discussion with Hannah Arendt (cf. Høibraaten 2001), which makes hope for counter-civil society courses possible in the first place. In this way, discourses can emerge from communicative practice and, in turn, affect and be taken up by the system level (cf. ibid, p. 160). Communicative power differs from domination or strategic communication by negotiating social rules and norms. "The agreement of those who deliberate together to act communally [...] signifies power insofar as it rests on conviction and hence on that peculiarly coercion-free force with which insights prevail." (Habermas 1983, p. 173) In a countermovement to Clausewitz's definition of war as a continuation of politics, Habermas defines "discourse as a continuation of communicative action by other means" (Habermas 1990, p. 130). Education is part of this discourse and can contribute to this discourse by reflecting on learning processes. This would be a response to Habermas' demand for an "attitude change accompanying the passage from communicative action to discourse" (ibid, p. 126).

Ultimately, however, this also means confronting those undesirable discourses from a pedagogical point of view, which claim exactly what, contrary to the educational goal, which is implicitly always present, from a peace science perspective. Changing from the system level to concrete conditions in the lifeworld can open up the pedagogical space of experience that frees from impotence. Here people act in dealing with the needs of their existence. Here they encounter impairments through dividing systems, and here they experience the disadvantage and the exclusion from the participation that would make them socially fully-fledged subjects in the first place. Nobody is only at home in the lifeworld since everyone is also integrated into systems. Nobody is only system representative because they are also in a lifeworld. This area of tension, in which the individual and society, biography and discourse are entangled, is equally the breeding ground for stereotyping and occlusion and opening and breaking stereotyping, depending on whether ambivalences are split off and suppressed or experienced and accepted as enriching.

Communication and Participation as Perspectives of Narrative and Participatory Pedagogy

A pedagogical silver bullet to counteract the division is the *com*munication and the corresponding *part*icipation. The pedagogical approaches to this exist. They are based on a narrative and participative exploration of life stories, life plans, and lifeworlds. These can be participatory projects in municipal settlements and neighborhoods, narrative groups in public and social institutions, performative explorations of living and working environments. Residents may photograph, film, document, and in this way rediscover their lives for themselves and others. City walks could invite people to explore public structures or discourse-building memorial sites. Educational theater projects such as forum theater (or theatre of the oppressed, Boal 1993) made accessible the themes of the lifeworld for reflection. The approaches are grounded on the assumption that blind adherence to the familiar can be opened up by processes of *Verwindung* (Vattimo 1987 referring to Nietzsche and Heidegger) in the meaning of distorted or twisted overcoming. A historical project about the mass emigration of South Tyrolean families "home to the Reich," adopted in the Hitler-Mussolini Agreement of 1939, may serve as an example. In this project, young people dealt with the migration trauma of their grandparents. They interviewed contemporary witnesses, collected memory objects, and performed the traumatic events artistically. The cultural memory (Assmann J. 2011) could be worked on a more fluid social memory, according to Aleida Assmann (2015), and helped the young people establish a new understanding of the current and fearful migration movements.

Learning *as* experience means that experiences do not obey interventions, nor are they open to direct access and insight, but they can be shared and co-experienced (cf. Laing 1977, p. 15). According to Dewey (1916/2009), experience only increases learning if we think from the solution found back to the problem that caused the learning experience (ibid, pp. 240–261). In this lies the educational hope of transformative learning, connecting the personal and the social level. Suppose problem-solving experiences living together in the lifeworld could be recovered and made fruitful for changing social narratives. In that case, this could affect the level of system and dominant practices.

The phenomenological attention for the potential of learning experiences in the lifeworld needs spaces for narrative exchange, participation, and reflexive appreciation of this experience. In this way, people's experiences in living together can enter into an individual and social consciousness and, as counterdiscourses of the lifeworld, enter into a dialogue with the dominant discourses of division.

Searching for the Lost Paradise

Educational Dilemmas and Potentials for a New Treatment of Nature and Earth

Given the adequately documented destruction and thinning of animal and plant life, the poisoning of air, earth, and water, does not require any particular evidence (cf. IPCC 2013; WWF 2016; Weizsäcker and Wijkman 2017). How this development, which encompasses all areas of life, progresses despite a broadly shared, generally accessible knowledge and makes political countermeasures appear helpless places educational science—together with all disciplines—under obligation. The compass of pedagogical attention is the evidence about the lost feeling for the connection between people and nature and earth.

There is a connection between the rejection of others because of their origin, language, skin color, different talents on the one hand, and the broken relationship of humans to nature on the other hand. It is the same psychodynamic motives that produce such attitudes. In the creation and combating of enemy concepts, as explained above, existential fears are discarded projectively and dumped on groups that facilitate such projections through specific characteristics. Discourses about identity, culture, and even *Heimat* show metaphorical traces originating from threats that humanity knows from its survival in nature. There is talk of being flooded by refugees, the flood of migration, wildfire, and foreign ulcers in our society. The equation of black with primordial nature shows the proximity of racism and the naturalization of the Other. Not least of all, pedagogy based its disciplinary and body-hostile concepts in the period after the first colonial expeditions on the equation of the *wild child* with the presumed *wild peoples* in the colonial territories. Just as colonization enslaved people not perceived as humans, the children had to be *made human* by strict discipline.

Thus, in racism, the dichotomous splitting towards nature and the animals as the supposedly wholly different ones works further on. Interestingly, the same populist movements that agitate against foreigners are also fighting against the

H. K. Peterlini, *Learning Diversity*,
https://doi.org/10.1007/978-3-658-40548-9_8

reappearance of wolves and bears in European countries and demand their re-extermination. At the same time, concepts of good respectful coexistence are rejected as impossible. These debates were particularly lively in South Tyrol in 2019, with advertising posters aimed at protecting the country from bear and wolf in almost the same tone as it was once defended from the Italian state. During the worldwide Covid 19 crisis, this media and political debate was temporarily silenced. Instead, there was now talk of a war against the virus, which we must win, whereas the virus had probably entered the human organism through human encroachments on the animal world. As soon as the first severe lockdown phase in the wake of the pandemic had subsided in May 2020, the debates on the shooting down of bears and wolves in South Tyrol were immediately resumed.

Of course, nature can be pretty threatening, just as it is the basis of the life to which we belong. The history of humankind is a history of inventions to assert oneself in a not always friendly nature and at the same time to wrest a maximum profit from it. In this way, the awareness has been lost that we are inextricably linked to nature, that humans and animals are common inhabitants of the earth, that we owe our food to the soil we exploit, and that we will have to continue breathing the air we consume and pollute.

For a diversity-educational approach in this subject area, too, it cannot be a matter of prescribing the correct behavior but instead dealing with this relationship between humankind and nature. The temptation would be great in this case to simply want to abolish the dichotomy that has arisen, which is probably difficult to do. However, we can make it a central theme and thus use it for educational processes.

This implies we must first consider the already discussed limits placed on education: by the indeterminable educational reception of every person, due to the teachers and educators' conditions, and through social and structural influences on every educational activity (cf. Bernfeld 1973). Education and training programs reach their limits where the efforts to "live the good life" (Vetter and Best 2015, p. 110) run counter to political, social, and economic decisions. Conversely, this does not mean that pedagogy can withdraw from its social and political responsibilities since politics—in the broadest sense—must not make pedagogy obsolete (cf. Hamburger 2010). According to Dietrich Benner, education is one of the constitutive human practices together with politics, economy, religion, ethics, and aesthetics. The task of pedagogy would be to transform the requirements of the other practices into "pedagogically legitimate" influences in a hierarchically non-subordinate discussion (Benner 2015, 61 ff). In this tension, the practice and science of education are in a difficult position, not only because the balance between the practices hardly corresponds to the non-hierarchical ideal relationship postulated by Benner. The influence on human action stands in the way of pedagogy álso due to dilemmas that need to be clarified.

The first dilemma is an epistemological inevitability. We have to objectify the matters we deal with, making them the object. That means: We have to build a distance between ourselves and them or at least imagine such a distance to be able to examine and discuss objects (in the broadest sense) as humans, to be able to think about them and be able to shape them. This applies to our understanding of nature, earth, and the environment, thus, how we deal with them.

The distance we need to think about nature, earth, and the world around us, to research it, to reflect on it, also creates the opportunity to exploit nature and earth as circumstances that have become foreign. We construct them fundamentally separate from us, as if we were not part of nature, as if we were not guests on this earth that supports and nourishes us. Our behavior on earth is as if we would not perish with it if we destroy nature—"like a face drawn in the sand at the edge of the sea," as Michel Foucault prophesied the idea of the human subject in the final sentence of "Order of Things" (Foucault 2002, p. 422).

Consciousness as the Fall—The Dividing of Man in Myth

The mental distance to earth, nature, and ourselves is necessary to define and control our relationship to the world, ecosystem, and us. At the same time, it creates the problems we want to handle. The distant perspective allows us to reflect and seemingly dominate the Other but veiled the interconnectedness of everything with everything—an ultimately indissoluble paradox. It is no coincidence that many myths of origin tell precisely that. We do not find the solution in these narrations. Still, the dilemma is fundamentally named, albeit—as it is peculiar to myths—in a transfigurative, distorted-healing way (cf. Freud 2011, p. 123). According to Waldenfels (2004a, p. 53), the productive achievement of the myth is that it "transforms the nothing (of which we know nothing) into something." The core of narration may remain nothing but helps find ground by narrating it. The most famous narration of origin in the European and Anglo-American region is expulsion from paradise. This story from the Bible tells of the beginning of human existence. It is also a metaphor for every person's birth in its life-creating and traumatizing scope (cf. Rank 1993).

Before the expulsion from paradise—into human existence, into life, into the world—Adam and Eve lived in divine bliss, nourished, born, relieved of all toil, with nature in peace and harmony (Gen. 2.1–2.25). Similarly, we can imagine everyone before their birth, in a state what we may later describe—as Freud called it—an "oceanic feeling" (Freud 1961, pp. 11–13) we will be missing forever and search, as discussed in the constructs of nation and *Heimat*, in unquestioned identity designs: protected, safe, nourished, borne, relieved of the hassle of individual and less secure identity designs and social disputes. What tore Adam and Eve from the unquestionable oneness with everything in the biblical myth was the reaching for the apple from the tree of knowledge. In German, it is the tree of *Erkenntnis*, which in its verb *Erkennen* is closer to cognition and recognition in the sense of recognizing someone and, in old usage, also stands for loving even in the sexual sense. In Hebrew, too, the biblical word for *Erkennen* in the purpose of physical love and procreation is the same as for understanding and grasping (Peterlini H.K. 2016a, p. 20). The fall from sin occurred simply as an increase in awareness, the cognitive revolution (cf. Harari 2014, pp. 3–83) with which Homo sapiens became aware of himself and thus became able to imagine (ibid, pp. 28–37).

A similar trace of the necessary splitting process is preserved in the German word for consciousness (*Bewusstsein*)—the prefix "-be" stands for duality. Awareness arises from separating the reflective subject from itself and thus from the world. Hegel attempted to grasp this need and necessity of division in the indissoluble dichotomy between lord and bondsman, according to which self-confidence must presuppose someone or something else to justify one's existence. At the same time, "its essential being is present to it in the form of an 'Other,' it is outside of itself and must rid itself of its self-externality" (Hegel 1977, p. 114). In Greek, Latin, Italian, and also in English, the etymology of consciousness shows a different trail: the prefixes *syn-* (synaísthēsis) and *con-* (conscientia, consapevolezza) refer to the connection that *con*sciousness has—across the division—to ourselves, to nature, to the earth by recognizing ourselves.

To recognize this grace of freedom is connected with the curse of expulsion from the paradisiacal world. At birth, the paradise of being lifted ends for every person; otherwise, we would suffocate in the womb. Birth gives us life and freedom also the hardships of existence. The word comes from Latin *ex-sistere* and means literally being outside of oneself. In the metaphorical narration, the first humans have to set off from a life without questions and worries towards pain and embarrassment, which triggers this knowledge of ourselves, of knowing oneself: "Then both eyes opened, and they realized that they were naked. They stitched fig leaves together and made an apron." (Gen 3,7) Then they hid under the trees. (Gen 3,8).

Plato's banquet tells the story of becoming human from another idea, namely the dichotomous division of the sexes. The spherical people represented a trinity (male, female, and both together), with four legs, four arms, and two faces. In their high spirits, an indirect parallel to the snake's promised empowerment of Adam and Eve through the forbidden apple, they became too powerful for the gods, so that Zeus split them in half (Plato 2008, p. 23). In this myth, a perfect and protected existence also ends with the punishment of division and exposition. The brutally divided two halves searched each other desperately but could only procreate into the earth until Zeus took pity and rebuilt their genitals in such a way that they at least could reunite. This myth, too, shows the interdependency of consciousness with procreation, albeit differently than in the biblical narrative: heterosexual love would henceforth serve to unite and procreate in the physical sense, while homoerotic male love, on the other hand, would create the beautiful and the true, ergo of knowledge (ibid, p. 24 ff.).

In both myths, the history of man begins with the loss of unity thought as natural. The act of separation begins the existence and the ability to recognize: consciousness. We have to take this human condition into account. As human beings, we are part of nature and alienated from it. We are guests on this earth, and through our cognitive distance, we have reached a dominant position, forgetting that we are equally dependent on nature and the world.

The Two Aspects of the Body—As Object to Control it, As Subject to Live with

The French philosopher René Descartes did not invent the split in our dual thinking, between *res cogitans* and *res extensa*, but described and established it powerfully. His name is in our intellectual history a synonym for the split between nature and reason, body and mind, body and soul. It seems our fate to be divided into two. Is there any escape? Phenomenology offers a fine distinction, in German between *Körper* and *Leib*, in English between *the body as object* and *body as the subject* or *lived body*. We can build up the same distance to the object-body as towards nature. We speak of parts of the body. We can explore the body like a thing, cure, operate, aestheticize, train, chemically manipulate it (cf. Böhme 2003, p. 12). The lived body eludes such access (ibid). We do not *have* it in the sense of possessing, but we *are body* as Maurice Merleau-Ponty (2002) condensed it: "I am not in front of my body, I am in it, or rather I am it." (Ibid, p. 173) And even more pointed: "I am my body." (Ibid, p. 231).

The two-body perceptions reflect the same tension between the understanding of nature-earth-man as a unity and the mastery of nature through distance. It enables humans to use (and exploit) the earth, to use (and exploit) animals and plants, as well as an understanding of body and mind that allows (at least supposed) control of the body by the mind. This talent (and curse) that we can distance ourselves mentally from what we are is at the root of causal thinking and its technical and natural sciences achievements. According to Böhme, we can only explore this through our distance to our body as if it were a foreign object, whereby we also forget that we are one as a body.

Humans cannot undo the path of separation, the exodus from the paradise of unquestionably being in nature. It may succeed in the mystical immersion or spiritual search, but this too must ultimately be initiated as a conscious exercise, as an active retreat, as a spiritual connection (*re-ligio*). The body of self-experience is accessible (Böhme 2003, p. 12), but not without the price of strangeness since we can only become aware of it by distancing the body (ibid, p. 34).

For pedagogy as a science of education, teaching, and accompaniment, there is an opportunity not to overcome the division but to address it and make it the subject of scientific research and criticism. Ultimately, there is no answer to the essence of the body because it eludes us as soon as we do not think of it as a body. The idea of the body as an animated body (Waldenfels 2000, p. 14; cf. Waldenfels 2004b) makes it possible to become aware of that other idea of the body as an object. The body thus serves as a concept for reflecting the division that separates us from ourselves and nature. At least the body concept can make us aware of it through this detour. If it is not already possible to overcome the division, we can use creative awareness of it. Its critical reflection opens up creative spaces for learning through and about the division.

For critical pedagogy on dichotomy, the body thus becomes a valuable figure of thought. Dichotomies are successful because they help open up objects, reduce complexity, simplify and rationalize processes, enable causal assumptions, allow categorizations, and thus also create social order. The price lies in eliminating the ambivalences, and the connections that exist despite the rational division between the parts split off. The problem is the inherent dynamism in the devaluation of one half in favor of the other. As soon as we think in dualisms, we assume an actual, "correct" half, the other half of which is only the deviation. We pose one half as dominant, while the other is hierarchically subordinate and inferior.

As discussed in previous chapters, the selectivity of dichotomous cleavages makes it possible for any connection between the two halves to be cut and made to disappear. Then the black man is so different from the white man that he can be enslaved and killed in the face of resistance that he is assigned a smaller compartment on the train that marriages between whites and blacks are prohibited. Then the woman is so utterly different from the man that her sexual desire can be denied. Her genitals may be mutilated, her access to education and political posts may be prohibited or at least limited, and she may be paid less. Then "the

Jew" is so different from the Aryan that he can be destroyed. Then nature, plants, and animals—although in the genetic code sometimes only minimally different from our own—are so different from humans that we can exploit, destroy, poison, extinguish them without thinking.

The construction of dichotomies, as it sounds in the Greek root words "δίχα/dícha" and "τέμνειν/témnein" for cutting in two, is a razor-sharp mental operation. It can make the split-off so strange that every connection is denied and thus also every sympathy and co-responsibility. These dynamics explain why political programs of democratically educated people can cooly argue why it is better to let refugees drown in the Mediterranean than let too many come to Europe. It will also be possible to terminate or sabotage the climate alliance or allow the import of elephant trophies to the United States again, thereby re-legitimizing a blind and ruthless big hunting game.

The Duplicated Heads of the Hydra—How to Deal with Dichotomies?

In the split subject, pedagogy goes into the very own field of the connecting, the intermediate. It is the pedagogical space for action where one half is deval-ued against the other. The dichotomy-critical attitude sharpens the awareness of inequality resulting from splitting-off-constructs. If, for example, school teaching reduces children to their brains as a cognitive machine, a body-phenomenological perspective brings the body as a neglected half and with that the child as a concrete human being into view. Knowledge in instruction often misses every liveliness and tangibility. The body-phenomenological perspective can inspire to involve nature, the earth, and life more in the classroom.

However, there is also a trap here: rehabilitating the devalued half of the dichotomous division alone would lead to the dichotomy trap again and quickly lead to the devaluation of the other half, such as framing cognitive learning and relying only on sensual comprehension. It's like the demonic snake from ancient Greece. If you cut off one head of the Hydra (Ogden 2013, 26 ff.), two new ones will grow from it. Dichotomies cannot simply be destroyed or overcome by reducing them to one side, as this would create new contradictions.

To put it bluntly, the point is not to change sides but to use the dichotomous field of tension for a pedagogy of experience: How do we experience ourselves as male and female and between the sexes? How do we experience gender at all? How do we experience rational thinking and intuitively acting people as body and mind and reason and soul? How do we experience ourselves as people in areas of tension between public and social inequality? Which separations can we track down, which connections can we trust without choosing one side or the other? It is an unfinishable exercise in the sense of a balanced dialectic without synthesis (cf. Meyer-Drawe 2000, pp. 63–73). For education approaches, this needs a sense of enduring ambiguity and ambivalence, according to which things are never unambiguous, never separate, but interwoven and intertwined.

Regarding the relationship between humans and the world: How do we experience nature, how do we experience animals, how do we experience the earth that supports us, how do we experience the asphalt that separates us from it and at the same time connects us with it? How do we experience the air we breathe in cities and on the mountain, the sun, the rain? How do we experience digging with our hands in the earth, milking a cow in the stable, riding a horse, how do we experience heat, cold, how do we experience water, how do we experience the scum in a sewage treatment plant? Knowledge is important, but for learning that brings or enables change, this experience level is indispensable.

This leads to an explicit pedagogical dilemma. Experience, as already discussed, can neither be instructed nor forced, neither pedagogically nor didactically. We can impart knowledge and check absorption and processing in educational processes. But whether children, adolescents, adults through this knowledge and beyond this knowledge, make learning experiences that let them deal more consciously with themselves, with their fellow human beings, with nature, with the earth, can neither be target-oriented guided nor precisely measured. Experience is pathic, in the sense of the ancient Greek *páthos* for experience, passion, suffering, even in the meaning of being passively exposed to it. This signifies that we cannot make experiences willingly. They happen to us, and they surprise us. They are *"Widerfahrnis,"* as it can be expressed more accurately in German, something that comes over us like astonishment. Astonishment surprises us. It is only accessible to us afterward. But we cannot decide to be astonished about something. So we cannot simply plan experiences through curricula, lesson preparations, education offers. It is simply not possible to determine what a child, an adolescent, an adult, or an aging person experiences or not in learning activities.

What we can and should do, is nevertheless not little. Education can create learning spaces for experiences and use the collected, discussed, perceived possible learning experiences for reflection on what has happened. It sounds simple but is seldom consequently practiced in educational contexts. Once again, let us refer to Dewey because it is a crucial understanding: According to Dewey (2009, pp. 240–261), people learn through experience, but this learning will not be fruitful if they do not reflect the gained knowledge. After having an aha-moment, it's essential to go back to the problem that has provoked and made possible the new understanding. So it is paramount not only to enable experiences but also to become aware of them and—again—to reflect on them.

Another dilemma: not only can experiences not be guided and determined in their implementation and progress, but they—unlike behavior—cannot be observed either. In this way, we humans ultimately remain blind to each other in our experiences. For Ronald D. Laing (1977), the existential necessity of an effort to understand the experiences of others arises: "I cannot avoid trying to understand your experience, because although I do not experience your experience, which is invisible to me (and non-tastable, non-touchable, non-smellable, and inaudible), yet I experience you as experiencing." (Ibid, p. 15 f).

Exploring Experiences: A Phenomenological Approach

Based on the described dilemmas, the phenomenological Vignette Research (Schratz et al. 2012; Agostini 2016; Baur and Peterlini 2016; Peterlini H.K. 2016a) tries to *co-experience* learning processes using a body-phenomenological research method. The researchers in the field pay attention to an attitude open to affection. They try to be aware of what happens bodily in learning situations, school events, and social interactions and become phenomenologically accessible. An example of a (slightly shortened) Vignette from the phenomenological oriented research project "Personal educational processes in heterogeneous groups"[1] (cf. ibid, pp. 9–14):

> Giselle, Gudrun, and Gitti are working on a worksheet on the topic of "dog." They also use a book about pets. Giselle happily says that she has a dog. She is the youngest in the group and keeps the book about pets with her at all times. When she's not using it, she puts her hand on it or presses it playfully against her stomach; otherwise, she flips through it, including when it comes to what is essential when you have a dog. Gudrun, meanwhile, already suggests: "Brush every day." Giselle flips through the book, apparently finds something, and reads it out loud to the other two: "regular care." [...] Then she reads out further keywords: "washing," "combing." Gudrun adds on her initiative: "Cutting nails!" Gitti comments: "Obviously, I knew that too." Gudrun answers Gitti slightly defiantly by pointing to Giselle: "I said it to her." Giselle continues reading from the book: "Doesn't cough." The three look at

---

[1] The phenomenological oriented "vignette research" started from the School of Education at the University of Innsbruck (headed by Michael Schratz). It has since spread to several university locations in Austria (Klagenfurt, Vienna), Italy, Germany, Switzerland, Greece, South Africa, and Vietnam (cf. https://vigna.univie.ac.at/en/).

each other. Giselle repeats: "The dog does not cough." They nod their nods and begin to write under "Characteristics": "The dog …" Since I am amazed, I ask the three: "Interesting, why is that so?" Gitti shrugs: "It has lungs." I answer: "We humans have those too, but we cough. You would have to consider why dogs do not cough." Giselle taps on her book, pointing to where she found the information: "It says: doesn't cough!" She now reads aloud: "Breathing … doesn't cough… it's here!" she repeats and passes it to me. I look at the position, notice the misunderstanding and give Giselle the book back with the question: "And what does it say over this sentence?" Now Giselle reads aloud: "Health characteristics…", pauses and laughs: "Oh … if he is healthy, he doesn't cough!" Now the other two are laughing too. Giselle puts the book down and begins to write. (Peterlini H.K. 2016a, p. 56).

Vignettes can be read from various perspectives; the interpretation is not about narrowing down to reconstructing the reality behind the written but rather about unfolding the possibilities of understanding what is shown in the Vignette. The question always is: What can we learn from this Vignette? The group work started with pleasure, the topic of dogs has a life-related connection, and Giselle has a dog at home. It is all the more interesting that this knowledge of everyday life is hardly mentioned, unlike the book that Giselle draws in and no longer leaves out, even presses it against her stomach. Does it give her security? There is a moment of upset in group work, but this gives way to astonishment when it comes to the dog not coughing. This claim amazes them, but: they believe the book. Even the question of the researcher, who wonders for himself, does not shake the knowledge of the book. What is black on white written there cannot be doubted. Giselle "knocks on her book." This symbolic gesture says: If it's in there, it has to be that way. As proof, she pushes the book to the researcher, who should see for himself.

When the misunderstanding clears up, everyone laughs. This laugh indicates that Giselle's insistence on the book was not shaken. Her source was not refuted; she had only misunderstood it. With a somewhat exaggerated comparison, Giselle could be a Bible expert who insists on the truth of a sentence that stands in black and white and therefore is not questioned by anyone, so weird, so illogical it may sound. Then a faded line above it is rediscovered on a parchment, reversing the sentence into its opposite. Nobody has to be ashamed, the book just did not reveal everything simultaneously, but it has retained its validity. The knowledge in the book is above the understanding of life experience. If it says that the dog doesn't cough, then it doesn't cough. (cf. ibid, pp. 57–58).

We all know textbooks and worksheets to prepare for exams, i.e., whit long lists of lichens divided into clear categories without the children have ever touched such a plant? The instruction tends not to make the learner familiar with lichens, how they feel, how they grow and smell; the task is to reproduce prefixed categories. Or worksheets on electrical energy without children experiencing a power surge or in the e-Factory to experience water violence? This criticism will not speak to a simple experiential pedagogy because even this cannot determine the experiences that learners have. However, an experience-oriented pedagogy that pays attention to corporeality could create opportunities for learning material and reflection on it, in the sense of *kairos*, an eye-opening moment.

Another Vignette from the project mentioned that such an experience—perhaps—was made

The children sit in a semicircle on the sofa in the animal corner, where animals have been kept in cages and showcased for some time, several types of insects, a guinea pig, stick snails, and hissing cockroaches. Mrs. Jennerwein, who stands in front of the children, shows them a dead insect. Most of the children wriggle their feet and lean forward. Jakob pushes himself entirely back into the sofa and looks at Mrs. Jennerwein unmoved. When Ms. Jennerwein asks, "Who wants to be the boss for the agate snails that we have just received," the children raise their arms one by one, and Jakob remains motionless. "Jan, then you look after them," Mrs. Jennerwein decides, "you have to make sure that they have enough moisture, enough food, and that everything is clean." Jan nods. Now those hissing cockroaches are allowed to circulate. Ms. Jennerwein explains that the animals hiss when they feel disturbed. Julia lets the cockroach crawl in the sleeves of her tight sweater, the boys next to her twist with laughter. Julia can hardly get the animal out of her sleeves. Ms. Jennerwein ends the laughter in a severe tone: "I want you to lose your fear, but keep the respect. If you are afraid, that is not a problem." Julia passes on the hissing cockroach with a redhead and holds her hand to her mouth to suppress the laugh. While the animal is being passed from hand to hand and hisses and hisses from time to time under the class's laughter, Ms. Jennerwein explains how the hissing cockroaches multiply. Next, a red-bellied toad can circulate. Ms. Jennerwein announces the animal as "skin-breathing, moist, slippery, of course, that can be a little…" Disgusted and sparkling again, the children quickly pass the little animal on, Julia reaches for it, but the toad hops past her to Jasmin, who passes it on to Jakob with an "Ätsch" call. The latter opens his hand and looks at the toad calmly, barely noticeably he smiles and says: "That's fine." (BJ1_11, cf. ibid, p. 65).

Whether Jakob has an experience that leads to learning, how deep the experience goes, what it can mean for him, can only be determined with the co-experience of the researcher condensed in the Vignette. The experience of experience requires the medium of narration, with which we can share experiences. Narrative, as a "given and take of talk",[2] is for Martin Buber (1937, p. 103) the medium of the between-human and therefore the educational "main portal" (ibid): "Here alone, then, as reality that cannot be lost, are gazing and being gazed upon, knowing and being known, loving and being loved." (Ibid).

The dichotomy between Jakob and an animal that causes disgust seems temporarily abolished for a short moment. Feeling the Other and feeling himself ("that's fine") is possible as a prerequisite for compassion. The Vignette tells about it. The reading of Vignettes at the level of scientific discourse and exchange with teachers is an open interpretation method. It helps to reflect on mostly unobserved magic moments in learning settings, often aside from the instruction, condensed in the short narration of Vignettes.

Empathy as Path to the Lost Half—The Encounter of Rosa Luxemburg

A touching document of compassion for an animal despite its own suffering is documented in the correspondence from Rosa Luxemburg. On Christmas Day 1917, she wrote a letter from the prison to the sister of comrade Karl Liebknecht. The woman describes how she was overwhelmed by pain and compassion when a soldier in the prison yard blew an emaciated buffalo until his skin tore: "I stood before it, and the beast looked at me, tears were running down my face— they were his tears." (Luxemburg 1921, p. 58) For Karl Kraus, this document of compassion for the suffering of an animal, which is representative of the abused being, was so moving and so instructive for a society of brutalization that he demanded its inclusion in the school curriculum (Kraus, 1920, pp. 5–9; Kraus and Luxemburg 2009).

It is noteworthy that Kraus refers to the above-discussed dilemma of educational thinking and acting, which consists of educational intentions' normative

---

[2] The original German version is based on the word sound of "Rede" (talk, speech) with "Redlichkeit" (honesty), which in the more recent translation by Walter Kaufmann (Buber 1970, p. 151) is translated as "honesty", with a reference to the multiple meanings in a footnote. Here, first translation by Ronald G. Smith ("talk") is preferred in order to emphasize the dialogical aspect, which is central to Buber and which would be reduced to an ethical aspect by the translation with honesty; this idea is also expressed in the German original version of "The Problem of Man", where narration is thought as the medium of the in-between, as the space of interpersonal experience (Buber 2001, p. 405). The two types of translation meet where Buber means the authentic speech, in which the interlocutors risk themselves.

setting. How can educators caught in the paradigms and thought patterns of their time presume to want to educate themselves for good, even if this good consists of making peace with nature and people (cf. Langer 1992)? Educating for good, for democracy, for peace involves the normative claim of knowing what is good. This ambition implies the dominant position of one who knows what others should think, how to live.

What children, what young people *should* or *have to* think about, can only be manipulated, if at all. To avoid that as possible, we have to balance the educational goal—which is implicit always—and the need to abstain from even subtle coercion. This balancing act may be possible if education enables experiences and accompanies reflections to relate people to themselves and their actions. Pedagogy can invite learners to enter the in-between and commune to earth, nature, animals, and plants, to others and strangers. Only there, in this in-between, is it possible to feel and empathize again. Empathy may be perhaps the only justifiable goal of pedagogical action. Encouraging people to relate to themselves and the Other (the nature, the stranger) maintained the necessary liberty of what people experience and how they reflect what is happening. It's a permanent invitation to step into the intermediate, connecting to the other dichotomous half in every topic. What this feeling and compassion accomplish is beyond the pedagogical approach. Nevertheless, it is difficult to imagine that compassion as feeling into others, strangers, animals, nature, bridging the abyss of division, could ever have negative consequences. "Passion commits no sin, only the cold" is a quote by the German writer Friedrich Hebbel (2017, p. 17).

Whether passion actually cannot sin is a matter of defining passion. Hebbel's diary entry challenged the strict and life-hostile morals of his time. In the cold, however, Adorno recognized one of the most profound causes for Auschwitz (Adorno 2010, p. 2). Suppose the path of becoming human, according to Horkheimer and Adorno (2002), has resulted in man-making and subjugating nature as the Other, including his drives, feelings, fears, longings. In that case, they see the only way out as "remembering nature in the subject" (ibid, p. 58) as a conscious act of rational and emotional re-appropriation of emotional rediscovery. The only thing that can prevent the destruction of everything we split off and consider alien from ourselves is active thinking and empathizing with the in-between. This opens the connection to the Other, nature, and ourselves as belonging to them through the separation. The thematization and critique of dichotomies are how pedagogy escapes its dilemmas and can build bridges across the division.

The experience-orientated pedagogical path makes diversity a richness rather than a loss-making deviation. And it valorizes our specificities rather than exposing them to discriminatory devaluation. Learning diversity is not so much a question of didactics, of something we find in this or that approach, in this or that pedagogical recipe block. Learning diversity means being involved in and reflecting on life experiences by examining one's patterns of perception and action. It includes one's stereotypes, one's dichotomies, one's fears, and closures, one's culturalisms, racisms, and one's indifference to others, whether they are human beings, animals, or the split-off world. It means a constant and never-ending attempt to search for the split-off half in our dichotomous thinking and living. There is no other way to try, again and again, to relate to the lost half when we lose ourselves in categorized visions of opinions, attitudes, situations, conflicts, problems, and other people. It is a learning that we have to practice daily and always throws us back to the beginning, where we will always fail a little bit. It is uncomfortable learning, which always obliges us to question the other side and ourselves. Even so, it is also enriching and liberating learning that opens our prisons of thought. Dichotomous patterns may be safeguarding and supporting, but much more, they are suffocating. By categorizing the world after dichotomous orders, we close ourselves in corsets.

Learning requires a risk. In learning experiences, we have to abandon the familiar by not yet having a guarantee about the new. Ultimately, all learning leads over this abyss. We have to release something (the familiar patterns, assumptions, knowledge, prejudices) and cannot yet rely on the new on the other side. Someone could say, why then should we learn? Why not stay how we are? It could seem simple to postulate that learning is, for humans, a principle of life, even there are good arguments for that. My answer would be yet more simple: We have no other chance. There is enough evidence of how the familiar has led humanity into permanent wars of all against all and into the extinction of the basis of life. We have to cross the abyss of the no-more and the not-yet: we have to learn. As a species that can intersubjectively share and reflect its experiences, we are gifted—and obliged—for transformative learning. Because, as Hannah Arendt has pointed out, we do not only move in circular paths but are also capable of transforming action. We can change our paths and patterns and re-learn and learn-anew. This gift is our responsibility, which we have towards ourselves, towards our fellow human beings, towards all fellow beings on earth, and towards the planet. The diversity that we would otherwise destroy is ourselves. Diversity is valued at all levels to be better understood and preserved.

Cool Aware Tutoolement releasment and taking each of the Creative Commons Attribution to international License, that Tutoolement manner not removed that which permits the use in an unrestricted with a copy/neproduction in any medium or format, as long as you give appropriate credit to the original author(s) and the source, provide a link to the Creative Commons license, and indicate if changes were made.

The images or other third material in this chapter are included in the chapter's Creative Commons license, unless indicated otherwise in a credit line to the material. If material is not included in the chapter's Creative Commons license and your intended use is not permitted by statutory regulation or exceeds the permitted use, you will need to obtain permission directly from the copyright holder.

On This Side of the Splitting Categories Lies the One World: Conclusions and Reflections for the Future of Education

This book has attempted a perhaps daring exploration. The ambivalences between belonging and differentiation, inclusion and exclusion, were explored first on the territorially and usually ethnically-nationally bound identity concept of *Heimat*. The following chapters gradually took up existential divisions in humanity and between humans, nature, and the world. Instead of comforting answers, the essays rather lead to new uncertainties, preferring ambivalences to unambiguity. Usually, we tend to resolve ambivalences to one side or the other and bring them into univocity. This would then mean rejecting *Heimat* as an exclusive concept tempting to nationalism or celebrating it as a blissful formula for sheltering, anchoring, assuring people in their social, territorial, and ecological lifeworld.

The Significance of Ambivalence for Pedagogy

The approach of this book is not to dissolve ambivalences but to look at them again and again in a suspended status. The effort is a phenomenology of diversity, which does not press life and its manifestations into categories. The ambition is not to bring them into an unambiguity but constantly subject them anew to contemplation, perception, analysis, and reflection. In this way, sometimes these, sometimes other aspects become more prominent. The object of investigation is not fixed but remains open for reconsideration in its development. In the distinction between purely creaturely life (*zoe*) and the social life of human beings (*bios*), deduced from Aristotle, the Italian philosopher Giorgio Agamben sees the root of all dichotomous divisions. It not only defines the clear separation of the human from animal and nature but reproduces itself within humanity as the distinction between humans with more and humans with fewer rights. As a kind of conception machine, an "outside" is produced by excluding something inside,

© The Author(s) 2023
H. K. Peterlini, *Learning Diversity*,
https://doi.org/10.1007/978-3-658-40548-9_9

although it would belong to it. Simply put, Humans exclude the animal from the totality of living beings just as they have excluded slaves, strangers, and currently refugees from humanity endowed with rights to treat them like animals (cf. Agamben 2004: p. 37). For undoing this mechanic dispositive, humanity would have to risk its confinement "in a suspension of suspension" and open the Sabbath—an allusion by Agamben to the Creation story—to human and animal (ibid: p. 92). *Suspension of suspension* also means leaving behind dialectical thinking, according to which two opposites fusion in a synthesis. Supposed that the formula man plus animal form the community of living beings, both man and animal would be living beings. The problem with this quite reasonable consideration is that a new objectification is created, suggesting new divisions.

Phenomenology tries to escape this—ultimately inescapable structure of human thinking—by a double negation. Let us play it out on the dispute if humans are self-determined (autonomous) or externally conditioned (dependent). The answer in the logic of double negation would be: Neither are we only autonomous nor only conditioned, neither only self-determined nor only externally determined. But rather: As concrete human beings, we find our ways by responding to the situations in which we are entangled. Our freedom does not arise in a vacuum but in responding to the living conditions to which we respond. If neither the one nor the other is assured, it remains and helps us only to deal with the concrete situation. Concerning the human-animal dilemma, it would sound like this: Neither is the human being only human nor the animal only animal because both are categories, not the concrete reality. As concrete beings, humans and animals constitute themselves in their life circumstances, their relations with each other and to each other, their questions of survival, their dependencies, and autonomous margins. Avoiding categorization helps to respond to what is concretely given in each case and not invoke the splitting order, for example, to let animals suffer for economic reasons or treat humans like animals. Then, the confrontation with the animal's suffering is required in the here and now and not in a definite order—the same concerns social disadvantages, discrimination against people with special needs, gender disequalities. Then categorical orders like middle class-working class, normal-disabled, man-woman may no longer legitimize their unequal treatment.

We know that the dichotomous machine is powerful. It works wonderfully, as if oiled: We owe to it the conquests of Classical mechanics and the social orders according to which our politics and our economy function. We owe it scientific access to the human body. We have made it an object (cf. Böhme 2003) and split it from our emotional constitution, the very body as *animate body* (Waldenfels 2000, p. 14) into bodily and mental.

Diversity-Reflective Education as Working with the Inevitable

At the end of this book, I do not want to fall into the illusion that we can escape dichotomies, that we can simply undermine them. Our thinking functions by distinguishing, and even this is a dichotomy since we separate thinking from feeling to speak of thinking. That our feeling affects how we think and our thoughts affect how we feel, is only half an understanding of the dichotomous problem. The deeper understanding would be to understand both as belonging together. Instead, we must constantly strive to think together something that only we have separated to use as a working concept of one and the other and, supposedly, get it under our dominion. Ultimately, the dichotomy is a tool for ordering the world that forces ourselves into that order. The device is not only passively involved in the whole process; on the contrary, it creates the object it is supposed to repair. In his critique of the power that the machine—as the totality of our orders— exerts on us, Ivan Illich saw only the chance that "we learn to invert the present deep structure of tools" (Illich 1973, p. 10), as mentioned above, by making the machine useful to us again to free ourselves from the enslavement of our own creatures (vulgo constructs).

How can this be done? To give a sure answer would be audacious. But if we know that we inevitably produce privilege and discrimination with distinctions, this must be countered by a permanent effort. It is the perception exercise as a pedagogical task (cf. Peterlini et al. 2020). We must always critically examine the instrument of dichotomy as we use it. We cannot do without distinctions because they are also helpful. If a child has special needs, it is necessary to recognize his impairment to support him. At the same time, however, we must also critically examine the tool we have used to categorize this child as having an impairment or learning difficulty: What has it shown us, and what is it obscuring? What is helpful, what is harmful about this categorization? What is beneficial, what is discriminatory? The exercise of perception requires looking at the child anew: What else is it besides weak in learning? What qualities do we miss when we identify it as having learning disabilities? This perception exercise requires a constant change of tools to bring the one clear distinction—weak in learning— back into a blur through other differences and perceptions. Then, perhaps, the child is freed from the categorical prison of "weak in learning," reborn into the child who is just sitting in front of us and whom I must engage with, regardless of how they are, but how they appear to me at the moment. Such engagement does not have to be a pedagogical idyll; it can still be complex and laborious. Still, it is the confrontation with life on this side of the category, the risk of relation with the concrete person or animal, the concrete refugee, the concrete homeless person, and the factual situation.

Learning diversity means exactly this—not a didactic technique, not a pedagogical recipe, but a constantly new engagement with diversity. In the pedagogical impetus, diversity is often presented as something beautiful to counter the negative connotations—such as mishmash, multiculturalism, loss of identity—with a positive spin. In fact, often, we may experience diversity beautifully and positively. But the danger is also that as a term with positive normative connotations, it becomes a euphemistic formula and often obscures the existential hardship associated with it. Diversity arises from differences; without differences, no diversity is conceivable. Differences, however, are double-edged. First, they simply mean that we are not all the same, but we differ. Waldenfels takes up Hegel's hypothesis that on the one hand, we need the Other to recognize ourselves, but that we thereby get outside ourselves. In this getting-out-of-itself, we ultimately reencounter ourselves because the foreign is us: "I get outside myself, not by chance, by illness, or out of weakness, but by being who I am" (Waldenfels 2004b, p. 242). The foreign, let us say the difference, is thus in ourselves, and only its acceptance opens the way to foreignness, to the difference of the others: "The foreignness in the midst of myself opens ways to the foreignness of the Other" (p. 244).

Dealing with Diversity—A Constant and Concrete Challenge

One considerable problem is that we are not constantly in the mode of philosophical meta-level where we see through such entanglements, but that we as human beings are at the mercy of that entanglement. Differences are not neutral. They are subject to discourses of power and help shape them; they can also constitute and protect privilege and be a projection surface for the discrimination through which they are produced in the first place. This concerns gender, age, language, origin, physical or health condition, socioeconomic and sociographic status (e. g., income, educational level, place of residence), and categories such as aptitude, disability, religious or ideological positioning, and many others. Each of these categories is intersectional overlapped and superimposition with other categories. This interconnectedness can strengthen or weaken positive or negative discrimination, knowing that positive discrimination can quickly turn negative.

To be pedagogically concerned with differences is itself an ambivalent endeavor. The difference at stake in the specific case can be reinforced and entrenched through the pedagogical concern. If it is downplayed, discrimination can continue to exist subliminally. Therefore, a diversity-sensitive and peace-oriented science of education must be aware of differences, without having a safe way how dealing with them at its disposal, but always trying to do so anew. There remains that risk of education as a relationship, which has already been addressed by Kupffer et al. (cf. 2000).

The challenge is first and foremost to perceive diversity on this side of dichotomous and mostly hierarchizing notions of normality. Categories of diversity are constructed and stabilized by specific marks like language, social and territorial origin, socioeconomic status, gender, skin color, sexual orientation, aptitude, disability, religious and ideological classification. "This side" means perceiving concrete persons in their responses to the world and history, even before they are categorized and diagnosed by isolated diversity characteristics. At the same time—and this is not a contradiction, but a challenging ambivalence—people often define themselves through those very differences which are the reason for their discrimination. This is not necessarily so but usually required to stand up for their rights, i.e., as a linguistic-ethnic minority, as disadvantaged gender, as a religious group, as diaspora community, as affected by social disability or socioeconomic inequality. In this dynamic lies the destructive and the productive potential of difference and diversity.

Diversity education thus stands in a tension between sharpening and unsharpening of difference, between appreciation and deconstruction. The interactions between the ethnicization of others and the ethnicization of oneself exemplify the double bind of denial and recognition. It reduces the subject to those aspects that are the motif of its discrimination, but through which it can also constitute. From this follows a pedagogical attitude that affirms differences and doubts them, takes them seriously, and relativizes them. Inequalities, globally and locally, in the classroom and social reality require a difference-reflective perception, without committing those affected to it, but seeing in their claims and contradictions the potential for empowerment and surpassing the self. From this positioning, let us return to the starting point of this book: *Heimat* can be the local familial environments, the regional landscape, and professional-private networking, or the orientation in a state entity with a national hegemonic effect.

It is not necessarily one or the other; it can be all or none of these. Whether people need a *Heimat* must not be discussed here. Undoubtedly, they need many of the circumstances readily associated with *Heimat*—belonging, security, social networking, and confirmation of their self-efficacy in the responsive events between people and the world. People need bearable living conditions that secure their existence, income, housing, and they need nature that makes life possible, i.e., air, water, life-enabling temperatures. It is not a very remote thought that these are universal needs. In many cases, their fulfillment in rich countries has been achieved by outsourcing the burdens and meeting one's own needs at the expense of others.

Well-being in Western countries has also been made possible by outsourcing social, economic, and environmental burdens (low wages, exploitation of human labor and natural resources, disposal of waste) and securing peace by outsourcing conflicts to proxy warriors in the externalization society (cf. Lessenich 2016). This is no longer possible in a globalized world—not only since Covid19. Flight movements, climate change, economic shifts of gravity, armament efforts of the hitherto excluded are indications we can no longer overlook. Simple, the small *Heimat* cannot be thought without its embedding in larger, even worldwide contexts. What is happening on the sizeable planetary level also causes problems on the individual and trim territorial levels. The division of the world into continental spheres of influence excludes those who belong to it and makes them foreign; what gets out of control in the process is the world itself.

"The Good Earth is Dying"—Approaches to Global Citizenship Education

The machine of division, described by Agamben, fails its service on the planetary level. There is no outside to which we can relocate our divisions in the future, at least no outside worth living in, no matter how often the super-rich would themselves shoot into space. Global citizenship in a homeland Earth, as Edgar Morin visionarily designed it, is neither a romantic globetrotting nor should it be political marketing. It is a necessity of life if we solve the problems we as humans are causing in the Anthropocene before they overwhelm us. Already 1971, Isaac Asimov wrote his expressive Essay "The good Earth is dying," asking for planetary strategies to save it:

> "The problems of our world are planetary. No nation is equal to them alone. However much individual nations may stabilize their populations within their means and protect their own environment, their efforts would remain futile if the rest of the world […] continued its poisoning activity. Even if each nation were to remedy the situation honestly but entirely on its own, the solutions of one nation would not necessarily match those of its neighbors, so that all efforts might fail. In short, problems of planetary scale require a planetary program and a planetary solution." (Asimov 1971).

Only one year later, the Club of Rome published its shocking report about the Limits of Growth (Meadows et al. 1972), which should awake the world, but remained simple an emphasized paper without concrete consequences. "Father, forgive them, for they do not know what they are doing" (Luke 23:34) are the words Jesus uses to excuse those who put him on the cross and share his garment. "For they do not know what they are doing" is the deduced German title for the famous James Dean movie "Rebel Without a Cause" (1955), the portrait of a lost generation of youth. The findings for the present ages should instead be the other way round: Why do we not what we know?

One possible answer could be the power of dichotomies that divides the world. Dichotomies cut the connection between the separate parts. If the Other is entirely different and the connection is lost, I can enslave it, exploit it, kill it, destroy it. Suppose the relation to other parts of the world is cut because of the sense of responsibility and connection end at national borders. In that case, we will hardly realize that fleeing people, climate change, economic and power unbalances, and even pandemics are like boomerangs of our own actions, our imperial lifestyle (cf. Wintersteiner 2021; Peterlini H.K. 2021b). And even if we know it, we will—blinded by the dichotomous perception—still not draw any consequence for our actions but cling to the illusion that the Flood is coming after us while it is already next door.

Thus, in addition to the blurring of differences, a second task arises for pedagogical action: to make the connection between the separated entities tangible again. If I experience that I always hurt myself in an argument, then I might be more likely to try to enter into a conversation. Suppose I can understand that my actions here cause suffering elsewhere, that I co-experience this suffering, and that it might come back to me. In that case, I might be more likely to change my behavior than if I am remain cut off from this possibility of co-experience.

We do not have certainty about this hope; it remains an attempt. In any case, the mediation of knowledge will not be enough; it requires establishing a relationship between the separate parts. I to connect to the difference in me and others, to animals, nature, and the problems in a faraway country. Dichotomous categorizations prevent this co-experience, prevent—in a word—empathy as compassionate connectedness. In a world divided by nations, hierarchies of power, economic interests, strengths of arms, disparities of wealth, empathy is also divided. I do not feel, or at least feel less, with those with whom I do not feel connected, with whom—supposedly—nothing connects me because they have been split off from me and placed in other categories. Suffering is less when it affects other countries and other social groups when it affects categories that—supposedly—have nothing to do with me because classified in other pigeonholes. We will only take responsibility for the Earth when we feel connected to it again when we experience that we are in the same drawer. Education cannot only convey this cognitively. It must make it possible to experience it. As discussed above, we need learning spaces where the encounter with the Other in a protected framework is relieved of its threatening nature. In such secure settings, free from competition and the logic of win or loos, learners may dare the discovery of one's own in the foreign, of the alien in one's own without danger.

For Transformative Learning Education has to Transform Itself

Let us open up a final thought. Can such diversity learning succeed in educational settings that are hierarchically structured, reproduce patriarchal patterns of domination, and in which learning is not appreciated as ongoing experience but is measured and evaluated by the output of predetermined performance categories? Such learning spaces are not experiential spaces in which trial and error remain harmless but are permeated by the dichotomous categories of right-wrong, reward-punishment, gifted-minor gifted, normal-disabled. According to Bourdieu (1977), they promote and reject children by the habitus they bring to school from unequal living conditions and carry them throughout their lives. A habitus-reflexive and thus difference-reflexive training for pedagogues in all contexts is a minimum requirement. Although a school conceived according to division structures, this too makes supportive handling of differences a pioneering achievement of the individual teachers, a miracle of education as a relationship that often succeeds against all odds. As precious the accomplishment of many teachers is, it often fails because of the structures that dictate the opposite. Hierarchy instead of shared discovery learning, performance output instead of process orientation, division according to categories gifted-minor gifted as a robust hidden curriculum that obscures and devalues the social differences and the different potentials of learners beyond the given categories.

Similarly, knowledge is divided into subjects and segregated by discourses of power. Schools and other educational settings underlie a counterproductive dynamic. They are asked to overcome the dichotomous orders to which they are subjected themselves. How should education free learners from binary thinking when wrong and right dominate almost every instruction? How can schools encourage students to venture beyond canonized knowledge if failure is punished with negative grades and possibly obstructed educational paths? How shall learners discover the connection between dichotomous separated parts when the learning stuff is itself divided artificially into traditional subjects. And finally: How can education alleviate the division of the world into national units and political-economic spheres of influence, when the monolingual habitus (Gogolin 2008) of the nation-state is politically imposed and pervades in part explicit, in part implicit the curricula.

There are significant claims to exact those questions by the latest strategy papers of the United Nations' educational institutions and the European Union. The Global Education Network Europe (GENE) is currently working on the further development of the Maastricht Declaration of 2002 into a "Declaration on Global Education to 2050" (GENE 2021). Already in the existing declaration, the goals are oriented towards a new approach to the division of the world:

> "Global education is education that opens people's eyes and minds to the realities of the globalized world and awakens them to bring about a world of greater justice, equity, and Human Rights for all. Global education is understood to encompass Development Education, Human Rights Education, Education for Sustainability, Education for Peace and Conflict Prevention and Intercultural Education; being the global dimension of Education for Citizenship." (Council of Europe 2002).

The Path for a new declaration, which is set to expire at the end of 2022, is justified by the need for broader and deeper political support for global education and concrete, practical steps.

Paper has been patient so far. This is also true for the UN Declaration on Global Citizenship Education, which can ultimately be an overall concept for difference-reflective education globally and locally. The program of UNESCO as the educational organization of the UN outlined in 2014, ratified by the member states in 2015 with the Agenda 2030, starts from the grave injustices and inequalities worldwide to see in Global Citizenship Education the response to these challenges:

> "While the world may be increasingly interconnected, human rights violations, inequality and poverty still threaten peace and sustainability. Global Citizenship Education (GCED) is UNESCO's response to these challenges. It works by empowering learners of all ages to understand that these are global, not local issues and to become active promoters of more peaceful, tolerant, inclusive, secure and sustainable societies." (United Nation 2015, Agenda 2030).

It's time for papers to get impatient, too, because the world's need and suffering can't wait any longer. UNESCO launched a broad process in 2021 to outline guidelines for Futures of Education. The drafts so far point out that individual measures and excellent schemes are not enough. Instead, there is a need for profound and structural transformations of schools. This means flattening the teacher-student hierarchy and a new understanding of knowledge and learning that no longer divides subjects by the hour, bud is developed in a participatory manner.

> "Pedagogy needs to move from a focus on teacher-driven lessons centered on individual accomplishment to instead emphasize cooperation, collaboration and solidarity. Curricula are often organized as a grid of subjects and need to shift to emphasize ecological, intercultural, and interdisciplinary learning. Teaching needs to move from being considered an individual practice to becoming further professionalized as a collaborative endeavor. Schools are necessary global institutions that need to be safeguarded. However, we should move from the imposition of universal models and reimagine schools, including architectures, spaces, times, timetables, and student groupings in diverse ways. In all times and spaces of learning we should move from thinking of education as mostly occurring in schools and at certain ages, and instead welcome and expand educational opportunities everywhere for everyone." (UNESCO 2021).

A new "social contract of education" is required to make learning a participatory adventure: "It is essential that everyone be able to participate in building the futures of education—children, youth, parents, teachers, researchers, activists, employers, cultural and religious leaders. We have deep, rich, and diverse cultural traditions to build upon. Humans have great collective agency, intelligence, and creativity. And we now face a serious choice: continue on an unsustainable path or radically change course." (Ibid).

The required planetary responsibility of education does not exclude the care for small homelands, just as the connectedness with others does not cause neglect of our own, but on the contrary, a rediscovery of ourselves. The earth is there wherever we put our feet on it, and wherever we shake hands with someone, we also feel ourselves.

Glossary

Names (pages not definitive—Seitenangaben nicht definitiv)

Abel, Andrea (p. 110, 203)

Adloff, Frank (p. 153, 203, 218)

Adorno, Theodor W. (p. 3, 6, 16, 121, 141–163, 177, 203, 204, 210)

Agamben, Giorgio (p. 181, 182, 186, 203)

Agostini, Evi (p. 173, 203)

Acciaioli, Greg (p. 216, 217)

Aigner, Josef Ch. (p. 20, 203)

Allolio-Näcke, Lars (2. 158, 203)

Améry, Jean (p. 11, 203)

Amplatz, Luis (p. 13)

Anzaldi, Linda (p. 124, 151, 203)

Arendt, Hannah (p. 118, 119, 161, 178, 204, 209, 210)

Arens, Susanne (p. 134, 159, 204)

Ariès, Philippe (p. 44, 204)

Asimov, Isaac (p. 186, 204)

Assmann, Aleida (p. 125, 145, 162, 204)

Assmann, Jan (p. 23, 145, 162, 204)

Ast, Gabriele (p. 19, 218)

Barthes, Roland (p. 9, 204)

Bauman, Zygmunt (p. 143, 152, 157. 20 4)

Baur, Siegfried (p. 11, 63, 173, 204)

Bauriedl, Thea (p. 122, 124148, 149, 150, 204)

Bausinger, Hermann (p. 12, 13, 19, 204)

Beilharz, Peter (p. 143, 204)

Benner, Dietrich (p. 166, 167, 204)

Berghold, Josef (p. 15, 22, 157, 204)
Bernfeld, Siegfried (p. 147, 148, 166, 205)
Best, Benjamin (p. 166, 218)
Bhabha, Homi (p. 37, 45, 112, 205)
Bienek, Horst (p. 11, 12, 205, 210)
Biewer, Gottfried (p. 132, 133, 205, 219)
Bion, Wilfried R. (p. 16, 205, 211)
Boal, Augusto (p. 162, 205)
Böhme, Gernot (p. 169, 170, 182, 205)
Böhnisch, Lothar (p. 44, 205)
Bourdieu, Pierre (p. 40, 110, 156, 188, 205)
Breinig, Helmbrecht (p. 2, 205)
Brunner, Claudia (p. 136, 205)
Buber, Martin (p. 121, 175, 176, 205)
Butler, Judith (p. 39, 42, 107, 205)
Canguilhem, George (p. 135, 205)
Chisholm, Lynne (p. 108, 206)
Gatterer, Claus (p. 26, 55, 208)
Clausewitz, Carl von (p. 161)
Deleuze, Gilles (p. 9, 206)
Delmonego, Ernst (p. 94, 206)
Demetrio, Duccio (p. 124, 203, 206)
Derrida, Jacques (p. 10, 21, 22, 206)
Descartes, René (p. 41, 169)
Dewey, John (p. 148, 162, 172, 206)
Diendorfer, Gertraud (p. 216)
Dietrich, Cornelie (p. 136, 206)
Eberle, Thomas (p. 135, 206)
Erdheim, Mario (p. 11, 19, 47, 135, 206)
Erikson, Erik H. (p. 34-37, 45, 46, 207, 211)
Fahrner, Eduard (p. 68)
Fairbairn, William R.D. (p. 15, 207)
Fanon, Frantz (p. 37, 38, 207)
Ferenczi, Sándor (p. 15, 207)
Feyerer, Ewald (p. 133, 208)
Fontana, Josef (p. 55, 207)
Foucault, Michel (p. 46, 134, 156, 167, 205, 207)
Freire, Paulo (p. 148, 207)

Key terms (pages not definitive—Seitenangaben nicht definitiv)

References

Abel, A., Ch. Vettori, and K. Wisniewski, Eds. 2012. *Kolipsi. Gli studenti altoatesini e la seconda lingua: indagine linguistica e psicosociale. Die Südtiroler SchülerInnen und die Zweitsprache: eine linguistische und sozialpsychologische Untersuchung*. Bozen-Bolzano: Eurac. http://www.eurac.edu/de/research/Publications/Pages/publicationdetails.aspx?pubId=0100156&type=Q. Accessed 15 May 2020.

Adloff, Frank. 2014. 'Wrong life can be lived rightly' Convivialism: Background to a Debate. Introduction". In *Convivialist Manifesto: A Declaration of Interdependence. French by Margaret Clarke*. Global Dialogue 3, 5–16. Duisburg: Käte Hamburger Kolleg, and Centre for Global Cooperation Research.

Adorno, Theodor W. 1955/2010. *Guilt and Defense. On the Legacies of National Socialism in Postwar Germany*. Edited and translated by Jeffrey K. Olick and Andrew J. Perrin. Cambridge: Harvard University Press.

Adorno, Theodor W. 1966/2010. Education After Auschwitz. In *Public Awareness Education Programs of the Sciences & Humanities – Technology & Global Bioethics*. Toronto: PAEP. https://docs.google.com/file/d/0ByVW1G--4tQDVFlLS3dyWHlON00/view. Accessed 20 June 2019.

Agamben, Giorgio. 2004. *The Open: Man and Animal*. Stanford: Standford University Press.

Agostini, Evi. 2016. *Lernen im Spannungsfeld von Finden und Erfinden. Zur schöpferischen Genese von Sinn im Vollzug der Erfahrung*. Paderborn: Schöningh.

Ahmed, Sara. 2000. *Strange Encounters. Embodied Others in Post-Coloniality*. London and New York: Routledge.

Ahmed, Sara. 2006. *Queer Phenomenology. Orientations, Objects, Others*. Durham: Duke University Press.

Aigner, Josef Ch. 2002. *Der ferne Vater. Zur Psychoanalyse von Vatererfahrung, männlicher Entwicklung und negativem Ödipuskomplex*. Gießen: Psychosozialverlag.

Allolio-Näcke, L., B. Kalscheuer, and A. Manzeschke, Eds. 2005. *Differenzen anders denken. Bausteine zu einer Kulturtheorie der Transdifferenz*. Frankfurt Main: Campus.

Améry, Jean. 1980. How much Home does a Person need? In *At the Mind's Limits: Contemplations by a Survivor of Auschwitz and Its Realities*, 41–61. Translated from German by Sidney and Stella P. Rosenfeld. Bloomington: Indiana University Press.

Anzaldi, L., D. Demetrio, and S. Rosatti. 1999. Un manifesto dell'educatore autobiografo. *Animazione sociale*, 29/3, 29–58.

© The Editor(s) (if applicable) and The Author(s) 2023
H. K. Peterlini, *Learning Diversity*,
https://doi.org/10.1007/978-3-658-40548-9

APA. 2016. Kurz setzt in Flüchtlingskrise auf Abschreckung. *APA Österreichische Nachrichtenagentur* 5 June 2016. http://www.nachrichten.at/nachrichten/politik/aussenpolitik/Kurz-setzt-in-Fluechtlingskrise-auf-Abschreckung;art391,2253223. Accessed 20 Oktober 2017.

Arendt, Hannah. 1958. *The Human Condition*. Chicago: University of Chicago Press.

Arendt, Hannah. 1960/1999. *Vita activa oder Vom tätigen Leben*. Munich, and Zurich: Piper.

Arens, S., and P. Mecheril. 2010. Schule – Vielfalt – Gerechtigkeit. Schlaglichter auf ein Spannungsverhältnis, das die erziehungswissenschaftliche Diskussion in Bewegung gebracht hat. *Lernende Schule* 13(49), 9–11.

Ariès, Philippe. 1962. *Centuries of Childhood. A Social History of Family Life*. Translated from the France by Robert Baldick. New York: Vintage.

Asimov, Isaac. 1971. The Good Earth is Dying, in *The Roving Mind*, 1983. Prometheus Books.

Assmann, Aleida. 2015. *Shadows of Trauma: Memory and the Politics of Postwar Identity*. Translated by Sarah Clift. Oxford: Oxford University Press.

Assmann, Jan. 2011. *Cultural Memory and Early Civilization: Writing, Remembrance, and Political Imagination*. Cambridge: Cambridge University Press.

Barthes, Roland. 1972. *Mythologies*. Translated from French by Annette Lavers. London: Paladin.

Bauman, Zygmunt. 2016. *Strangers at Our Door*. Cambridge: Polity Press.

Bauman, Zygmunt. 2014/2017. "Zuckerberg ha puntato con Facebook su paura solitudine", interview on the occasion of the presentation of the Hemingway Prize 2014 in Lignano (Italy). *Ansa* January 2017. http://www.ansa.it/sito/notizie/cultura/2014/06/27/bauman-zuckerberg-ha-puntato-con-facebook-su-paura-solitudine_1f0e937a-1bb5-4168-be1c-104241e864f4.html. Accessed 15 January 2020.

Baur, Siegfried. 2000. *Die Tücken der Nähe. Kommunikation und Kooperation in Mehrheits-/Minderheitensituationen*. Meran-Merano: Alpha Beta.

Baur, S., I. von Guggenberg, and D. Larcher. 1998. *Zwischen Herkunft und Zukunft. Südtirol im Spannungsfeld zwischen ethnischer und postnationaler Gesellschaftskultur. Ein Forschungsbericht*. Meran-Merano: Alpha Beta.

Baur, S., and H.K. Peterlini, Eds. 2016. *An der Seite des Lernens. Erfahrungsprotokolle aus dem Unterricht an Südtiroler Schulen – ein Forschungsbericht*. Erfahrungsorientierte Bildungsforschung Vol. 2. Innsbruck, Vienna, Bozen-Bolzano: Studienverlag.

Bauriedl, Thea. 1988. *Das Leben riskieren. Psychoanalytische Perspektiven des politischen Widerstands*. Munich, Zurich: Piper.

Bausinger, Hermann. 1980. *Heimat* und Identität. In *Heimat. Sehnsucht nach Identität*. Ed. Elisabeth Moosmann, 30–71. Berlin: Ästhetik und Kommunikation.

Bausinger, Hermann. 2000. *Typisch deutsch. Wie deutsch sind die Deutschen?* Munich: C.H. Beck.

Beilharz, Peter. 2000. *Zygmunt Bauman: Dialectic of Modernity*. London: Sage.

Benner, Dietrich. 2015. *Allgemeine Pädagogik. Eine systematisch-problemgeschichtliche Einführung in die Grundstruktur pädagogischen Denkens und Handelns*. Weinheim: Juventa.

Berghold, Josef. 2005. *Feindbilder und Verständigung. Grundfragen der politischen Psychologie*. Wiesbaden: VS.

Bernfeld, Siegfried. 1925/1973. *Sisyphus or the Limits of Education.* Foreword by Anna Freud, Preface by Peter Paret, Translated by Frederic Lilge. Berkeley, Los Angeles, London: University of California Press.

Bhabha, Homi. 1983. The Other Question ... Homi K. Bhabha Reconsiders the Stereotype and Colonial Discourse. *Screen* 24/6, 18–36.

Bhabha, Homi. 1996. Culture's In-Between. In: *Question of Cultural Identity*, Eds. S. Hall, and P. Du Gay, 53–60. London, Thousand Oaks, and New Delhi: Sage.

Bienek, Horst. 1985. Vorbemerkung des Herausgebers. Warum dieses Buch? In *Heimat: neue Erkundungen eines alten Themas*, Ed. Horst Bienek, 7–8. Munich: Hanser.

Biewer, Gottfried. 2010. *Grundlagen der Heilpädagogik und Inklusiven Pädagogik.* Bad Heilbrunn: Klinkhardt.

Bion, Wilfried R. 1963. *Elements of Psychoanalysis.* London: William Heinemann.

BMBWF. 2018. *Deutschförderklassen und Deutschförderkurse. Pressegespräch mit Bundesminister Univ.-Prof. Dr. Faßmann 17.4.2018.* https://www.bmbwf.gv.at/dam/jcr:891 4959b-54fb-4fa8-ac7d-1ccbcde0a245/180416_Pr%C3%A4sentation_PK_Deutschf% C3%B6rderklassen.pdf. Accessed 20 april 2020.

Boal, Augusto. 1993. *Theatre of the Oppressed.* New York: Theatre Communications Group.

Böhme, Gernot. 2003. *Leibsein als Aufgabe. Leibphilosophie in pragmatischer Hinsicht.* Zug: Die graue Edition.

Böhnisch, Lothar. 2008. Lebenslage Jugend, sozialer Wandel und Partizipation von Jugendlichen. In *Jugendliche planen und gestalten Lebenswelten. Partizipation als Antwort auf den gesellschaftlichen Wandel.* Eds. Th. Ködelpeter, and U. Nitschke, 25–40. Wiesbaden: VS.

Bourdieu, Pierre. 1977. *Outline of a Theory of Practice.* Cambridge: Cambridge University Press.

Bourdieu, Pierre. 2002. *Masculine Domination.* Translated from French by Richard Nice. Stanford: Stanford University Press.

Breinig, H., and K. Lösch. 2006. Transdifference. *Journal for the Study of British Cultures* 13/2, 105–122.

Buber, Martin. 1937. *I and Thou.* Translated from German by Ronald Gregor Smith. Edingburgh: T. & T. Clark.

Buber, Martin. 1970. *I and Thou.* Translated from German by Walter Kaufmann. New York: Charles Scribner's Sons.

Buber, Martin. 2001. *Das Problem des Menschen.* Munich: Gütersloher Verlagshaus.

Bundeskanzleramt. 2017. *Zusammen für Österreich. Regierungsprogramm 2017–2022.* https://www.bundeskanzleramt.gv.at/documents/131008/569203/Regierungsprogramm_ 2017%E2%80%932022.pdf/b2fe3f65-5a04-47b6-913d-2fe512ff4ce6. Not accessible any more, see FPÖ 2017.

Brunner, C. 2021. *Epistemische Gewalt Wissen und Herrschaft in der kolonialen Moderne.* Bielefeld: Transcript. DOI: https://doi.org/10.14361/9783839451311

Butler, Judith. 1993. *Bodies That Matter: On the Discursive Limits of "Sex".* London: Routledge.

Butler, H., and G. Ch. Spivak. 2007. *Who Sings the Nation-state? Language, Politics, Belonging.* Calcutta: Seagull Books.

Canguilhem, George. 1991. *The Normal and the Pathological.* Translated by Carolyn R. Fawcett in collaboration with Robert S. Cohen, with an introduction by Michel Foucault. New York: Zone Books.

Council of Europe. 2002. *The Maastricht Global Education Declaration: Achieving the Millennium Goals.* https://rm.coe.int/168070e540. Accessed 10 M

Chisholm, L., and H.K. Peterlini. 2012. *Aschenputtels Schuh. Jugend und Interkulturelle Kompetenz in Südtirol-Alto Adige. Forschungsbericht über einen verkannten Reichtum.* Meran-Merano: Alpha Beta.

Deleuze, Gilles. 2003. *Desert Islands and Other Texts, 1953–1974.* Los Angeles: Semiotext(e).

Delmonego, Ernst. 1988. *Lüsen: Natur – Kultur – Leben.* Lüsen: Gemeinde Lüsen.

Demetrio, Duccio. 1998. *Pedagogia della memoria. Per se stessi, con gli altri.* Rome: Meltemi.

Derrida, Jacques. 1973. *Speech and Phenomena and other Essays on Husserl's Theory of Signs.* Evanston: Northwestern University Press.

Derrida, Jacques. 1997. *Of Grammatology.* Translated by Gayatri Chakravorty Spivak. Baltimore: Johns Hopkins University Press.

Der Spiegel. 1984. Sehnsucht nach *Heimat. Der Spiegel* 40, 252–263.

Der Spiegel. 2012. Was ist *Heimat*? *Der Spiegel* 15, 60–71.

Dewey, John. 1916/2009. *Democracy and Education: An Introduction to the Philosophy of Education.* Waiheke Island: Floating Press.

Dietrich, Cornelie. 2017. Im Schatten des Vielfaltsdiskurses: Homogenität als kulturelle Fiktion und empirische Herausforderung. In *Differenz – Ungleichheit – Erziehungswissenschaft. Verhältnisbestimmungen im (Inter)Disziplinären,* Eds. I. Diehm, M. Kuhn, and C. Machold, 123–138. Wiesbaden: Springer.

Die Presse. 2005. Was heißt *Heimat*? I–II. *Spectrum – Die Presse,* 14 May 2005.

Die Zeit. 2017. Flüchtlinge: Ohne Schutz gegen Gleichgültigkeit und Kälte. *Die Zeit,* 11 January 2017. http://www.zeit.de/gesellschaft/2017-01/fluechtlinge-winter-griechenland-balkanroute-fs. Accessed 20 January 2017.

Dolomiten. 2006a. Kameradinnen in Rock oder Lederhosen. *Dolomiten.Tagblatt der Südtiroler* 25 August 2006, 15.

Dolomiten. 2006b. Zwischen Schnaps und Gewehr. *Dolomiten. Tageblatt der Südtiroler* 26/27 August 2006, 15.

Dolomiten. 2009. "Wo hört 'wünschenswerte Heimatverbundenheit' auf, wo fängt Rechtsextremismus an". *Dolomiten. Tagblatt der Südtiroler* 12 November 2009, 90.

Du. 2009. Heimat auf Zeit. Vom Leben im Anderswo. *Du 802 (monothematic).*

Duden, Eds. 2007. *Das Herkunftswörterbuch. Etymologie der deutschen Sprache.* Vol. 7. Mannheim-Leipzig-Vienna-Zurich: Dudenverlag.

Eberle, Thomas. 2009. Ethnomethodologie. In: *Qualitative Marktforschung. Konzepte – Methoden – Analysen,* Eds. R. Buber and H. H. Holzmüller, 93–109. Wiesbaden: Gabler Verlag.

Ein Tirol. 2013. Der Identitätsverlust hält auch bei den Schützen Einzug. *Informationsseite: Ein Tirol* 28 November 2013. https://www.facebook.com/eintirol/posts/der-identit%C3%A4tsverlust-h%C3%A4lt-auch-bei/598587096857596/. Accessed 20 September 2018.

EQF. 2008. *The European Qualifications Framework.* https://europa.eu/europass/en/european-qualifications-framework-eqf. Accessed 10 January 2021.

Erdheim, Mario. 1984. *Die gesellschaftliche Produktion von Unbewusstheit. Eine Einführung in den ethnopsychoanalytischen Prozess.* Frankfurt Main: Suhrkamp.

Erikson, Erik H. 1945. Childhood and Tradition in Two American Indian Tribes. A Comparative Abstract, with Conclusions. In *The Psychoanalytic Study of the Child. The official Journal for the Association for Child Psychoanalysis,* Vol. 1, 319–350.

Erikson, Erik H. 1950. *Childhood and Society.* New York: WW Norton & Co.

Erikson, Erik H. 1964/1994. Identity and Up-rootedness in Our Time. In *Erik H. Erikson, Insight and Responsibility. Lectures on the Ethical Implications of Psychoanalytic Insights,* 81–107. New York: WW Norton & Co.

Erikson, Erik H. 1968. *Identity. Youth and Crisis.* New York: W. W. Norton & Co.

Erikson, Erik H. 1971. *Einsicht und Verantwortung. Die Rolle des Ethischen in der Psychoanalyse.* Frankfurt Main: Fischer.

Erikson, Erik H. 1974. *Dimension of a New Identity: The 1973 Jefferson Lectures in the Humanities.* New York: W. W. Norton & Co.

Erikson, Erik H. 1994. *Identity and the Life Cycle.* New York: W. W. Norton & Co.

Fairbairn, William R.D. 1952. *Psychoanalytic studies of personality.* London: Tavistock Publications with Routledge & Kegan Paul.

Fanon, Frantz. 1967. *Black Skin, White Masks.* Translation from French by Charles Lam Markmann. New York: Grove Press.

Ferenczi, Sándor. 1949. Confusion of the Tongues Between the Adults and the Child. *International Journal of Psycho-Analysis* 30 (4): 225–230.

Fontana, Josef. 1993. *Neumarkt 1848–1970: Ein Beitrag zur Zeitgeschichte des Unterlandes.* Bozen-Bolzano: Athesia.

Foucault, Michel. 1986. *Of Other Spaces: Utopias and Heterotopias.* Translated from the French by Jay Miskowiec. *Diacritics* 16, 22–27.

Foucault, Michel. 1995. *Discipline and Punish. The Birth of Prison.* Translated from French by Alan Sheridan. New York: Vintage.

Foucault, Michel. 2002. *Order of Things.* London-New York: Routledge.

FPÖ. 2017. *Zusammen für Österreich. Regierungsprogramm 2017–2022.* https://www.fpoe.at/fileadmin/user_upload/www.fpoe.at/dokumente/2017/Zusammen_Fuer_Oesterreich_Regierungsprogramm.pdf. Accessed 20 June 2020.

Freiheitsliebe. 2016. *Norbert Hofer: „Deine Heimat braucht dich jetzt", Deutscher!* https://diefreiheitsliebe.de/politik/norbert-hofer-deine-Heimat-braucht-dich-jetzt-deutscher. Accessed 8 June 2019.

Freire, Paolo. 2007. *Pedagogy of the Oppressed.* New York: Continuum.

Freud, Sigmund. 1900/1913. *The interpretation of dreams.* Translated by Abraham A. Brill. New York: Macmillan Company.

Freud, Sigmund (1905/1962): *Three Essays on the Theory of Sexuality.* Translated from German by James Strachey. New York: Basic Books.

Freud, Sigmund. 1911/1958. *Psycho-Analytic Notes on an Autobiographical Account of a Case of Paranoia (Dementia Paranoides)* [Case of Schreber]. Edited and translated from German by James Strachey. Standard Edition Vol. XII., 1–82. London: Hogart Press.

Freud, Sigmund. 1914/1975. On narcissism: an Introduction. In *Sigmund Freud, On the History of the Psycho-Analytic Movement, Papers on Metapsychology and Other Works.* Edited and translated from the German by James Strachey, Standard Edition Vol. XIV, 103–142. New York: Vintage

Freud, Sigmund. 1915/1975. The Unconscious. In *Sigmund Freud, On the History of the Psycho-Analytic Movement, Papers on Metapsychology and Other Works*. Edited and translated from the German by James Strachey, Standard Edition Vol. XIV, 150–216. New York: Vintage.

Freud, Sigmund. 1922/2011. *Group Psychology and the Analysis of the ego*. Edited and translated from German by James Strachey. London: The international Psychoanalitical Press.

Freud, Sigmund. 1923/1975. The Ego and the Id. In *Sigmund Freud, The Ego and the Id and Other Works*. Edited and translated from German by James Strachey, Standard Edition Vol. XIX, 12–68. New York: Vintage.

Freud, Sigmund- 1930/1961. *Civilization and its Discontents*. Translated and edited by James Strachey. New York: W. W. Norton.

Frieters-Reermann, Norbert. 2016. Friedenspädagogik als zivile Konfliktbearbeitung. Spannungsfelder ziviler Friedensbildung. In *Friedenspädagogik und Demokratiepädagogik. Jahrbuch Demokratiepädagogik 2016/17*, Eds. H. Rademacher and W. Wintersteiner, 50–68. Schwalbach/Ts.: Wochenschau.

Feyerer, E., and A. Holzinger. 2018. Inklusive Bildung. Die erziehungswissenschaftliche Antwort auf die Diversität der Gesellschaft. In *Baustellen in der Bildungslandschaft. Zum 80. Geburtstag von Peter Posch,* Eds. Herbert Altrichter et al, 204–215. Münster: Waxmann.

Garfinkel, Harold. 1967. *Studies in Ethnomethodology*. Cambridge: Polity Press.

Gatterer, Claus. 2003. *Schöne Welt, böse Leut. Kindheit in Südtirol*. Vienna: Folio.

Geertz, Clifford. 1973. Thick Description: Towards an Interpretive Theory of Culture, In: *Clifford Geertz, The Interpretation of Cultures*, 311–323. New York: Basic Books.

Geisen, Thomas. 2018. Rassismus als Herausforderung für die Soziale Arbeit in der Migrationsgesellschaft. In *Migration und Soziale Arbeit,* 40, 100–106.

GENE. 2021. *A new European Declaration on Global Education to 2050*. https://www.gene.eu/ge2050. Accessed 18 December 2021.

Geo. *Heimat*. Warum der Mensch sie wieder braucht. *Geo Magazin* 10, 103–165.

Gnutzmann, Claus. 2004. Mehrsprachigkeit als übergeordnetes Lernziel des Sprach(en)unterrichts: die 'neue' kommunikative Kompetenz? In *Mehrsprachigkeit im Fokus: Arbeitspapiere der 24. Frühjahrskonferenz zur Erforschung des Fremdsprachenunterrichts,* Eds. K.-R. Bausch, F. G. Königs, and H.-J. Krumm, 45–54. Tübingen: Günther Narr.

Goethe, Johann Wolfgang von. 1789. The Goodlike. In *The Library of the World's Best Literature. An Anthology in Thirty Volumes*, C.D. Warner et al., first published 1917. https://www.bartleby.com/library/poem/2238.html. Accessed 19 december 2019.

Goethe, Johann Wolfgang von. 1814/1962. Sprichwörtlich. In *Berliner Ausgabe: Poetische Werke 1–16*, Vol. 1, 431–463. Berlin: Aufbau-Verlag.

Gogolin, Ingrid. 2008. *Der monolinguale Habitus der multilingualen Schule*. Internationale Hochschulschriften Vol I. Münster: Waxmann.

Golowitsch, Helmuth. 2009. *Für die Heimat kein Opfer zu schwer. Folter – Tod – Erniedrigung: Südtirol 1961–1969*. Nürnberg: Verlag Kienesberger E.

Graham, Sharyn. 2002. Sex, Gender, and Priests in South Sulawesi, Indonesia. In *IIAS-Newsletter, Research & Reports* 29/11, 27.

Gramsci, Antonio. 1928–1937/1971. *Selection from the Prison Notebooks*, Eds. Q. Hoare and G.N. Smith. London: Lawrence and Wishart.

Gstettner, Peter. 1981. *Die Eroberung des Kindes durch die Wissenschaft. Aus der Geschichte der Disziplinierung*. Reinbeck bei Hamburg: Rowohlt.

Habermas, Jürgen. 1971. *Knowledge and Human Interests*, translated from German by Jeremy J. Shapiro. Boston: Beacon Press.

Habermas, Jürgen. 1979a. Moral Development and Ego Identity. In: *Communication and the Evolution of Society*, translated from German by Thomas McCarthy, 69–94. Boston: Beacon Press.

Habermas, Jürgen. 1979b. Historical Materialism and the Development of Normative Structures. In *Communication and the Evolution of Society*, translated from German by Thomas McCarthy. 95–129. Boston: Beacon Press.

Habermas, Jürgen. 1979c. Towards a Reconstruction of Historical Materialism. In: *Communication and the Evolution of Society*, translated from German by Thomas McCarthy, 130–177. Boston: Beacon Press.

Habermas, Jürgen. 1983. Hannah Arendt: On the Concept of Power (1976). In *Philosophical-political Profiles Studies in Contemporary German Social Thought*, edited and translated from German by Thomas McCarthy, 171–188. Cambridge: MIT Press.

Habermas, Jürgen. 1984. *Theory of Communicative Action. Vol. 1: Reason and Rationalisation of Society*. Translated from German by Thomas McCarthy. Boston: Beacon Press.

Habermas, Jürgen. 1987. *The Theory of Communicative Action, Vol. 2: A Critique of Functionalist Reason*. Translated from German by Thomas McCarthy. Boston: Beacon Press.

Habermas, Jürgen. 1990. *Moral Consciousness and Communicative Action Studies*. Translated by Christian Lenhardt and Shierry Weber Nicholsen; introduction by Thomas McCarthy. Cambridge: MIT Press.

Habermas, Jürgen. 2009. *Europe: The Faltering Project*. Cambridge: Polity Press.

Halbwachs, Maurice. 1980. *The collective memory*. New York: Harper & Row.

Hall, Stuart. 1990. Cultural identity and diaspora. In *Identity: community, culture, difference*, Ed. Jonathan Rutherford, 223–237. London: Lawrence & Wishart.

Hall, Stuart. 1996. New ethnicities. In *Critical Dialogues in Cultural Studies*, Eds. D. Morley and K.-H. Chen, 441–451. London: Routledge.

Hamburger, Franz. 2010. Über die Unmöglichkeit, Politik durch Pädagogik zu ersetzen. In *Bei Vielfalt Chancengleichheit. Interkulturelle Pädagogik und Durchgängige Sprachbildung*, Eds. M. Krüger-Potratz, U. Neumann, H. Reich, 16–23. Münster: Waxmann.

Hahn, H. H., and E. Hahn. 2002. Nationale Stereotypen. Plädoyer für eine historische Stereotypenforschung. In *Stereotyp, Identität und Geschichte. Die Funktion von Stereotypen in gesellschaftlichen Diskursen*, Eds. H.H. Hahn and E. Hahn with the assistence by Stephan Scholz, 19–56. Frankfurt Main, Berlin, Bern, Bruxelles, New York, Oxford, Wien: Peter Lang.

Han, Byung-Chul. 2013. *Digitale Rationalität und das Ende des kommunikativen Handelns*. Berlin: Matthes & Seitz.

Han, Byung-Chul. 2017. *The Expulsion of the Other: Society, Perception and Communication Today*. Translated from German by Wieland Hoban. Cambridge: Polity Press.

Harari, Yuval N. 2014. *Sapiens: a brief history of humankind*. Translated from Hebrew by the author with the help of John Purcell and Haim Watzman. London: Vintage Book.

Hebbel, Friedrich.1903/2017. *Tagebücher 1: 1835–1839. Neue historisch-kritische Ausgabe*, Ed. Monika Ritzer. Berlin: De Gruyter.

Hegel, Georg W. F. 1807/1977. *Phenomenology of spirit*. Translated from German by A. V. Miller with Analysis of the Text and Foreword by J. N. Findlay. New York and Oxford: Oxford University Press.

Heidegger, Martin. 1927/1962. *Being and Time*. Translated from German by J. Macquarrie and E. Robinson. Oxford: Basil Blackwell.

Heinz, W., G. Kayser, and E. Knödler-Bunte. 1980. Sehnsucht nach Identität. Schwierigkeiten mit *Heimat* von links her umzugehen. In *Heimat. Sehnsucht nach Identität*, Ed. Elisabeth Moosmann, 30–49. Berlin: Ästhetik & Kommunikation.

Heitmeyer, Wilhelm. 1994. *Das Gewalt-Dilemma*. Frankfurt Main: Suhrkamp.

Hill, M., and E. Yıldız. 2018. Einleitung. In *Postmigrantische Visionen. Erfahrungen – Ideen – Reflexionen*, Eds. M. Hill, and E. Yildiz, 7–9. Bielefeld: Transcript.

Hofer, Christine. 2010. *Die pädagogische Anthropologie Maria Montessoris – oder: Die Erziehung zum neuen Menschen*. Würzburg: Ergon.

Høibraaten, Helge. 2001. Kommunikative und sanktionsgestützte Macht bei Jürgen Habermas mit einem Seitenblick auf Hannah Arendt. In „*The Angel of History is looking back*". *Hannah Arendts Werk unter politischem, ästhetischem und historischem Aspekt. Texte des Trondheimer Arendt-Symposions vom Herbst 2000*, Eds. B. Neumann, H. Mahrdt, and M. Frank, 153–194. Würzburg: Königshausen & Neumann.

Honneth, Alex. 1995. *The Struggle for Recognition: The Moral Grammar of Social Conflicts*. Cambridge: Polity Press.

Horkheimer, M., and Th. W. Adorno. 1944/2002. *Dialectic of Enlightenment*. Translated from German by Edmund Jephcott. Brooklyn: Stanford University Press.

Humboldt, Wilhelm von. 1783/1980. Theorie der Bildung des Menschen. In *Wilhelm von Humboldt, Werke in fünf Bänden*, Vol 1, Eds A. Flitner und K. Giel, 56–233. Darmstadt: Wissenschaftliche Buchgesellschaft.

Husserl, Edmund. 1913/1990: *The Idea of Phenomenology*. Translated by W. P. Alston and G. Nakhnikian. Boston and London: Kluwer Academic Publisher.

Huxel, Katrin. 2014. *Männlichkeit, Ethnizität und Jugend. Präsentationen von Zugehörigkeit im Feld Schule*. Wiesbaden: Springer.

Illich, Ivan. 1973. *Tools for Conviviality*. London: Calder & Boyars.

Illich, Ivan. 2002. *Deschooling Society. London: Publisher Marion Boyars.*

IPCC. 2013. Working Group I: Contribution to the Fifth Assessment Report of the Intergovernmental Panel on Climate Change. In *The Physical Science Basis. Summary for Policymakers*, Eds. Th. F. Stocker, D. Qin, G.-K. Plattner, M. M.B. Tignor, S. K. Allen, J. Boschung, A. Nauels, Y. Xia, V. Bex, P. M. Midgley. http://www.climatechange2013. org/images/report/WG1AR5_SummaryVolume_FINAL.pdf. Accessed 3 January 2018.

Jäger, Uli. 2016. Friedensbildung und -pädagogik. Strukturelle Verankerungen und Initiierung von Lernprozessen. In *Friedenspädagogik und Demokratiepädagogik. Jahrbuch Demokratiepädagogik 2016/17*, Eds. H. Rademacher und W. Wintersteiner, 21–30. Schwalbach/Ts.: Wochenschau.

Jens, Walter. 1985. Nachdenken über *Heimat*. Fremde und Zuhause im Spiegel deutscher Poesie. In *Heimat: neue Erkundungen eines alten Themas*, Ed. Horst Bienek, 14–26. Munich: Hanser.

Jungwirth, Ingrid. 2007. *Zum Identitätsdiskurs in den Sozialwissenschaften. Eine postkolonial und queer informierte Kritik an George H. Mead, Erik H. Erikson und Ervin Goffman*. Bielefeld: Transcript.

Kernberg, Otto F. 1975. *Borderline conditions and pathological narcissism*. New York: Aronson.

Klein, Melanie. 1946. Notes on some schizoid mechanisms. *International Journal of Psychoanalysis* 27, 99–110.

Klein, M., and J. Riviere. 1953. *Love, Hate, and Reparation*. Psycho-Analytical Epitomes 2. New Impression. London: Hogarth Press and the Institute of Psycho-Analysis.

Kraus, Karl. 1920. Der tiefste, je in einem Saal … Mit einem Brief von Rosa Luxemburg an Sophie Liebknecht vom Dezember 1917. *Die Fackel* 546, 5–9.

Kraus, K., and R. Luxemburg. 2009. *Büffelhaut und Kreatur: die Zerstörung der Natur und das Mitleiden des Satirikers*, Ed. Friedrich Pfäfflin. Berlin: Friedenauer Presse.

Krone. 2016. Türkei rüstet sich. Apokalypse in Aleppo: Weitere Million Flüchtlinge? *Krone* 5 February 2016. http://www.krone.at/welt/apokalypse-in-aleppo-weitere-million-fluech tlinge-tuerkei-ruestet-sich-story-494660. Accessed 15 June 2016.

Kuckartz, Udo. 1998. *Umweltbewusstsein und Umweltverhalten*. Berlin: Springer.

Kupffer, Heinrich. 1984. *Der Faschismus und das Menschenbild in der deutschen Pädagogik*. Frankfurt Main: Fischer.

Kupffer, H., J. Schiedeck, D. Sinhart-Pallin, and M. Stahlmann, Eds. 2000. *Erziehung als offene Geschichte. Vom Wissen, Sprechen, Handeln und Hoffen in der Erziehung*. Weinheim: Beltz und Deutscher Studienverlag.

Kurier. 2016. FPÖ-Plakat: „Aufstehen für Österreich". *Kurier* 14 March 2016. https://kurier. at/politik/inland/bp-wahl-fpoe-plakat-ruft-zum-aufstehen-fuer-oesterreich-mit-norbert-hofer-auf/186.808.292. Accessed 30 April 2017. Accessed 12 October 2019.

Kurier. 2017a. Landespolizeidirektor Pilsl: „Wir wollen nichts beschönigen". *Kurier* 1 January 2017. https://kurier.at/chronik/oberoesterreich/landespolizeidirektor-pilsl-wir-wollen-nichts-beschoenigen/238.676.447. Accessed 8 January 2017a.

Kurier. 2017b. Analyse: Weniger Menschen kehren freiwillig zurück. *Kurier* 10 May 2017. https://kurier.at/chronik/oesterreich/warum-es-mehr-abschiebungen-gibt/262.906.726. Accessed 10 January 2017b.

Kurier. 2018. Deutschklassen: Das sind die ersten Entwürfe für die Lehrpläne. *Kurier* 31 July 2018. https://kurier.at/politik/inland/deutschklassen-das-sind-die-ersten-entwuerfe-fuer-die-lehrplaene/400076453. Accessed 30 August 2018.

Lahme-Gronostaj, Hildegard. 2003. „Trauma" und „katastrophische Veränderung" (Wilfred Bion). In *Trauma, Beziehung und soziale Realität*, Eds. M. Leuzinger-Bohleber, and R. Zwiebel, 61–60. Tübingen: Edition diskord.

Laing, Ronald D. 1977. *The Politics of Experience and The Bird of Paradise*. Harmondsworth: Penguin.

Lakitsch, M., and A.M. Steiner, Eds. 2015. *Gewalt für den Frieden? Vom Umgang mit der Rechtfertigung militärischer Intervention*. Vienna-Berlin: Lit-Verlag.

Lang, Claudia. 2006. *Intersexualität: Menschen zwischen den Geschlechtern*. Frankfurt Main: Campus.

Langer, Alexander. 1992. *Vie di pace – Frieden schließen. Rapporto dall'Europa – Berichte aus Europa.* Mezzocorona. Edizioni Arcobaleno.

Langer, Alexander. 2015. Zehn Punkte fürs Zusammenleben. In *Jenseits von Kain und Abel. Zehn Punkte fürs Zusammenlesen – neu gelesen und kommentiert. In Memoriam Alexander Langer 1995–2015*, Eds. M. Boschi, A. Jabbar, and H.K. Peterlini, 7–18. Meran-Merano, and Klagenfurt-Celovec: Alpha Beta und Drava.

Larcher, Dietmar. 2000. Barbie und der Terminator lernen die Demokratie. Überlegungen zur pädagogischen Subversion patriarchaler Geschlechtsrollenklischees. In *Männlichkeit und Gewalt. Konzepte für die Jugendarbeit*, Eds. I. Bieringer, W. Buchacher, E. J. Forster, 46–54. Opladen: Leske + Budrich.

Larcher, Dietmar. 2005. *Heimat* – Eine Schiefheilung. Südtirols große Erzählungen. Ein Versuch der Dekonstruktion. In *Fremdgehen. Fallgeschichten zum Heimatbegriff*, Eds. D. Larcher, W. Schautzer, M. Thuswald, 165–195. Meran-Merano and Klagenfurt-Celovec: Alpha Beta and Drava.

Lassnig, Ewald. 2012. *Dorfbuch der Gemeinde Partschins: mit den Ortsteilen Partschins, Rabland, Töll, Quadrat, Vertigen, Tabland, Sonnenberg.* Partschins: Gemeinde Partschins.

Leiprecht, Rudolf. 1992. *Unter Anderen – Rassismus und Jugendarbeit.* Duisburg: Institut für Sprach- und Sozialforschung.

Lessenich, Stephan. 2016. Living Beyond the Means of Others. In *The Futures We Want: Global Sociology and the Struggles for a Better World*, Ed. Markus S. Schulz (ed.), 66–67. Berlin: ITF.

Liddell, H.G., and R. Scott. 1940. *A Greek–English Lexicon.* Oxford: Clarendon Press. http://www.perseus.tufts.edu/hopper/text?doc=Perseus:text:1999.04.0057:entry=sxolh/. Accessed 14 Juny 2020.

Loose, Rainer. 1997. *Prad am Stilfserjoch: Beiträge zur Orts- und Heimatkunde von Prad, Agums und Lichtenberg im Vinschgau/Südtirol.* Edited by Marktgemeinde Prad am Stilfserjoch. Lana: Tappeiner.

Luhmann, Niklas. 1981. Unverständliche Wissenschaft. Probleme einer theorieeigenen Sprache. In *Soziologische Aufklärung 3. Soziale Systeme, Gesellschaft, Organisation*, 170–177. Opladen: Westdeutscher Verlag.

Luhmann, N., and K.-E. Schorr. 1982, Das Technologiedefizit der Erziehung und die Pädagogik. In *Zwischen Technologie und Selbstreferenz. Fragen an die Pädagogik*, 11–40. Frankfurt Main: Suhrkamp.

Lutz, H., and N. Wenning, Eds. 2001. *Unterschiedlich verschieden. Differenz in der Erziehungswissenschaft.* Opladen: Leske + Budrich.

Luxemburg, Rosa. 1921. *Letters from Prison: with a Portrait and a Facsimile.* Berlin: Publishing House of the Young International.

McLuhan, Marshall. 1962. *The Gutenberg Galaxy: the making of typographic man.* Toronto: University of Toronto Press.

Mead, George H. 1934. *Mind, Self and Society from the Standpoint of a Social Behaviorist.* Chicago: Chicago University Press.

Mead, Margaret. 1967. *Male and Female: A Study of the Sexes in a Changing World*, New York: William Morrow & Company.

Mead, Margaret. 2001. *Sex and Temperament: In Three Primitive Societies.* New York: Harper Perennial.

Meadows, D.H., D.L. Meadows, J. Randers. W.W. Behrens III. 1972. *The Limits to Growth. A report to the Club of Rome's Project on the Predicament of Mankind.* Washington DC: Potomac Associates – Universe Books.

Mecheril, Paul. 2002. Natio-kulturelle Mitgliedschaft – ein Begriff und die Methode seiner Generierung. In *Tertium comparationis* 8/2, pp. 104–115.

Merico, Maurizio. 2007. *Il tempo in frammenti. Giovani, tempo libero e consumo.* Lecce: Kurumuny.

Merleau-Ponty, Maurice. 1945/2002. *Phenomenology of Perception.* Translated from French by Colin Smith. London and New York: Routledge.

Merz, C., and H. Qualtinger. 1996. Der Herr Karl. In *Helmut Qualtinger, „Der Herr Karl" und andere Texte fürs Theater,* workedition Vol. 1, Ed. Traugott Kritschke, 173–175. Vienna: Deuticke.

Meyer-Drawe, Käte. 1986. Lernen als Umlernen – Zur Negativität des Lernprozesses. In *Lernen und seine Horizonte. Phänomenologische Konzeptionen menschlichen Lernens – didaktische Konsequenzen.* Eds. W. Lippitz and K. Meyer-Drawe, 19–45. Königstein/Ts.: Cornelsen.

Meyer-Drawe, Käte. 1987. Die Belehrbarkeit des Lehrenden durch den Lernenden – Fragen an den Primat des Pädagogischen Bezugs. *In Kind und Welt. Phänomenologische Studien zur Pädagogik,* Eds. W. Lippitz, and K. Meyer-Drawe, 63–73. Frankfurt Main: Athenäum.

Meyer-Drawe, Käte. 2000. *Illusionen von Autonomie. Diesseits von Ohnmacht und Allmacht des Ich.* Munich: P. Kirchheim.

Meyer-Drawe, Käte. 2003. Lernen als Erfahrung. *Zeitschrift für Erziehungswissenschaft 6(4),* 505–514.

Meyer-Drawe, Käte. 2018. *Die Welt als Kulisse. Übertreibungen in Richtung Wahrheit,* Ed. Nordrhein-Westfälischen Akademie der Wissenschaften und der Künste. Paderborn: Ferdinand Schöningh.

Mitscherlich, Alexander. 1969. *Society without the Father.* Translated from German by Eric Mosbacher. London: Tavistock Publications.

Mitterhofer, S., and G. Obwegs, Eds. 2000. *Es blieb kein anderer Weg... Zeitzeugenberichte und Dokumente aus dem Südtiroler Freiheitskampf.* Auer: Arkadia.

Morin, E., and A. B. Kern. 1999. *Homeland Earth.* New York: Hampton Press.

Müller, Romy. 2018. „Deutsch-Mandarin, Syrisch-Deutsch oder Slowenisch-Deutsch?, *ad astra* 6/1, 6–10.

Nancy, Jean-Luc. 2008. *Corpus.* Translated from French by Richard A. Rand. New York: Fordham University Press.

Niethammer, Lutz. 2000. *Kollektive Identität. Heimliche Quellen einer unheimlichen Konjunktur.* Reinbek: Rowohlt.

NTZ. 2006a. Die unangenehme Wahrheit. *Neue Südtiroler Tageszeitung* 26/27 June, 4.

NTZ. 2006b. „Mit der Zeit gehen". *Neue Südtiroler Tageszeitung* 26/27 June 2006, 4.

Oe24. 2016. 10.000 müssen heuer gehen – Bald mehr Härte bei Abschiebungen. *Oe 24,* 27 November 2016. http://www.oe24.at/oesterreich/politik/Bald-mehr-Haerte-bei-Abschiebungen/260188361. Accessed 10 January 2017.

Ogden, Daniel. 2013. *Drakon: Dragon Myth and Serpent Cult in the Greek and Roman Worlds.* Oxford: Oxford University Press.

Oxford Dictionaries. 2016. *Word of the year is ...* https://en.oxforddictionaries.com/word-of-the-year/word-of-the-year-2016. Accessed 28 December 2016.

Parin, Paul. 1999. „Die Weißen denken zuviel". Über das Eigene und das Fremde – im Gespräch mit Paul Parin. In *Die Geschichte ist nicht zu Ende! Gespräche über die Zukunft des Menschen und Europas,* Ed. Hans-Jürgen Heinrichs, 163–179. Vienna: Passagen Verlag.

Peterlini, Hans Karl. 1992. *Bomben aus zweiter Hand. Zwischen Gladio und Stasi: Südtirols missbrauchter Terrorismus.* Bozen-Bolzano: Raetia.

Peterlini, Hans Karl. 1998. Fesselnde *Heimat.* Südtirol, das Entstehen einer Verteidigungskultur. In *Sehnsucht Heimat. Katalog zur Ausstellung im Salzlager Hall,* Ed. Benedikt Erhard, 74–89. Hall: Kunsthalle Tirol.

Peterlini, Hans Karl. 2000. *Heimat*graffiti. Die Farbenlehre der Tiroler. In *A. Kofler, and H.K. Peterlini, Graffiti in Tirol – Graffiti in Tyrol,* Ed. Durst Phototechnik, 104–107. Innsbruck: Haymon.

Peterlini, Hans Karl. 2010a. "*Heimat*"-Homeland between life world and defence psychosis. Intercultural learning and unlearning in an ethnocentric culture: long-term study on the identity formation of junior "Schützen" (shooters). In *Procedia – Social and Behavioral Sciences 5,* 59–68. https://doi.org/10.1016/j.sbspro.2010.07.051.

Peterlini, Hans Karl. 2010b. *Freiheitskämpfer auf der Couch. Psychoanalyse der Tiroler Verteidigungskultur von 1809 zum Südtirol-Konflikt.* Innsbruck: Studienverlag.

Peterlini, Hans Karl. 2011. *Heimat zwischen Lebenswelt und Verteidigungspsychose. Politische Identitätsbildung am Beispiel junger Südtiroler Schützen und Marketenderinnen.* Innsbruck: Studienverlag.

Peterlini, Hans Karl. 2012. *Capire l'altro. Piccoli racconti per fare memoria sociale.* Milan: Franco Angeli.

Peterlini, Hans Karl. 2015a. Werkzeuge der Konvivialität. Friedensarbeit im Zeitalter von Entgrenzung, Migration und neuen Bedürfnissen nach geschützten *Heimat*en. In *Jenseits von Kain und Abel. Zehn Punkte fürs Zusammenlesen – neu gelesen und kommentiert. In Memoriam. Alexander Langer 1995–2015,* Eds. M. Boschi, A. Jabbar, and H.K. Peterlini, 37–54. Meran-Merano, and Klagenfurt-Celovec: Alpha Beta and Drava.

Peterlini, Hans Karl. 2016a. *Lernen und Macht. Prozesse der Bildung zwischen Autonomie und Abhängigkeit.* Innsbruck: Studienverlag.

Peterlini, Hans Karl. 2016b. Vom Stammeln zum Sprechen. Sprachpolitik, Ethnizität und Migration im Spannungsfeld zwischen nationaler Dominanz und Ermächtigung zur Teilhabe. In *Jenseits der Sprachmauer. Erinnern und Sprechen von Mehrheiten und Minderheiten in der Migrationsgesellschaft,* Ed. Hans Karl Peterlini, 143–171. Klagenfurt-Celovec and Meran-Merano: Drava and Alpha Beta.

Peterlini, Hans Karl. 2016c. *100 Jahre Südtirol. Die Geschichte eines jungen Landes.* Innsbruck, Vienna, and Bozen-Bolzano: Haymon.

Peterlini, Hans Karl. 2017a. Erziehung nach Aleppo. Pädagogische Reflexionen zu Rechtspopulismus, Rassismus und institutioneller Kälte gegenüber Menschen in Not. In *Migration. Bildung. Frieden. Perspektiven für das Zusammenleben in der Postmigrantischen Gesellschaft,* Eds. B. Gruber, and V. Ratković, 175–200. Münster: Waxmann.

Peterlini, Hans Karl. 2017b. Between Stigma and Self-Assertion: Difference and Belonging in the Contested Area of Migration and Ethnicity. In *A Land on the Threshold. South Tyrolean Transformations 1915–2015,* Eds. G. Grote, and H. Obermair, 341–360 Bern: Peter Lang.

Peterlini, Hans Karl. 2018. Die Normalisierung des Anders-Seins Phänomenologische Unter-richtsvignetten und Reflexionen zur gelebten Inklusion im italienischen Schulsystem am Beispiel von Südtiroler Schulen. In *Zeitschrift für Inklusion* 1, https://www.inklusion-onl ine.net/index.php/inklusion-online/article/view/406 vom 8.2.2018.

Peterlini, Hans Karl. 2019a. Falsche Kinder in der richtigen Schule – oder umgekehrt? Auslo-tungen eines Perspektivenwechsels von selektiven Normalitätsvorstellungen hin zu einer Phänomenologie des 'So-Seins'. In *Ist inklusive Schule möglich? Nationale und Interna-tionale Perspektiven,* Eds. E. J. Donlic, E. Jaksche-Hoffman, and H.K. Peterlini, 41–60. Bielefeld: Transcript.

Peterlini, Hans Karl. 2019b. Über den Abgrund der Dichotomie. Pädagogische Dilemmata und Perspektiven für einen neuen Umgang mit Natur und Erde. In *Io corpo – Io racconto – Io emozione,* Ed. L. Dozza, 31–43. Bergamo: Zeroseiup.

Peterlini, Hans Karl. 2020a. „Die nicht wissen, wo sie hingehören." Jugendliche Identitäts-bildung im Kontext von Ethnisierung. In *Pädagogische Anthropologie der Jugendlichen,* Eds. S. Blumenthal, S. Sting, and J. Zirfas, 198–215. Weinheim: Beltz Juventa.

Peterlini, Hans Karl. 2020b. Die Übung des Wahrnehmens als pädagogische Aufgabe – Ein-führung und Vorwort. In *Wahrnehmung als pädagogische Übung. Theoretische und prax-isorientierte Auslotungen einer phänomenologisch orientierten Bildungsforschung,* Eds. H.K. Peterlini, I. Cennamo, J. Donlic, 7–10. Innsbruck-Vienna-Bozen: Studienverlag.

Peterlini, Hans Karl. 2021a. *Feuernacht. Südtirols Bombenjahre. Hintergründe – Schicksale – Bewertungen.* Third edition. Bozen-Bolzano: Raetia.

Peterlini, Hans Karl. 2021b. Warum...? Und wenn ja, wie anders? Pädagogische Antwortver-suche auf politische und gesellschaftliche Überlebensfragen. In *W. Wintersteiner, Die Welt neu denken lernen – Plädoyer für eine planetare Politik. Lehren aus Corona und anderen existentiellen Krisen,* Ed. H.K. Peterlini, 11–28. Bielefeld: transcript.

Peterlini, H.K., I. Cennamo, and J. Donlic, Eds. 2020. *Wahrnehmung als pädagogische Übung. Theoretische und praxisorientierte Auslotung einer phänomenologisch orien-tierten Bildungsforschung.* Innsbruck, Vienna, and Bozen-Bolzano: Studienverlag.

Peterlini, Julia. 2015. Das Recht auf 'effektiven' Unterricht in den Regelklassen von Men-schen mit Behinderung und dessen Verwirklichung in Italien und Südtirol. In *RdJB* 64 (1), 90–104.

Peterlini, Oskar. 2009. The South-Tyrol Autonomy in Italy, Historical, Political and Legal Aspects. In *One Country, Two Systems, Three Legal Orders – Perspectives of Evolution,* Eds. J. Oliveira, and P. Cardinal, 143–170. Berlin: Springer.

Petri, Rolf. 2001. Deutsche *Heimat* 1850–1950. Konsum und Region im 20. Jahrhun-dert. In *Comparativ. Leipziger Beiträge zur Universalgeschichte und vergleichenden Gesellschaftsforschung* 11/1, 77–127.

Pistolato, Franco. 2016. Friedensforschung in Italien? Nicht wirklich, aber ... In *Friedens-forschung in Österreich. Bilanz und Perspektiven, Jahrbuch Friedenskultur 2015,* Eds. W. Wintersteiner, and L. Wolf, 155–167.

Plato. 2008. *The Symposium.* Edited by M. C. Howatson and Frisbee C. C. Sheffield, Trans-lated by M. C. Howatson.Cambridge: Cambridge University Press.

Pörksen, Uwe. 1995. *Plastic Words: The Tyranny of a Modular Language.* Philadelphia: Pennsylvania University Press.

Purschert, Patricia. 2012. Postkoloniale Philosophie. Die westliche Denkgeschichte gegen den Strich lesen. In *Schlüsselwerke der Postcolonial Studies,* Eds. J. Reuter, and A. Karentzos, 243–354, Wiesbaden: Springer VS.

Radtke, Frank-Olaf. 2013. *Vom Multikulturalismus zur Parallelgesellschaft – Selbstvergewisserung in der Einwanderungsgesellschaft.* https://Heimatkunde.boell.de/2006/09/18/vom-multikulturalismus-zur-parallelgesellschaft-selbstvergewisserung-der. Accessed 14 December 2019.

Rank, Otto. 1924/1993. *The Trauma of Birth.* Edited with a new introduction by James Lieberman. New York: Dover Publications.

Reheis, Fritz. 2016. „Friede den Hütten, Krieg den Palästen!". Herausforderungen für die Friedenspädagogik im 21. Jahrhundert. In *Friedenspädagogik und Demokratiepädagogik. Jahrbuch Demokratiepädagogik 2016/17,* Eds. H. Rademacher and W. Wintersteiner, 31–43, Schwalbach/Ts., Wochenschau.

Reitmair-Juárez, Susanne. 2016. Entwicklungen, Schwerpunkte und Methoden der Friedenspädagogik. In *Friedensforschung, Konfliktforschung, Demokratieforschung. Ein Handbuch,* Eds. G. Diendorfer, B. Bellak, A. Pelinka, and W. Wintersteiner, 180–214. Cologne, Weimar, and Vienna: Böhlau.

Reitz, Edgar. 2010. *Heimat. Hunsrücktriologie,* http://www.Heimat123.de. Accessed 20 July 2010.

Ribolits, Erich. 1997. *Die Arbeit hoch? Berufspädagogische Streitschrift wider die Totalverzweckung des Menschen im Post-Fordismus.* Munich and Vienna: Profil.

Richter, Horst E. 1972. *Die Gruppe. Hoffnung auf einen neuen Weg, sich selbst und andere zu befreien. Psychoanalyse in Kooperation mit Gruppeninitiativen.* Reinbek-Hamburg: Rowohlt.

Römhild, Regina. 2014. Jenseits ethnischer Grenzen. Für eine postmigrantische Kultur- und Gesellschaftsforschung. In *Nach der Migration. Postmigrantische Perspektiven jenseits der Parallelgesellschaft,* Eds. E. Yildiz, and M. Hill, 37–48. Bielefeld: Transcript.

Röttger-Rössler, Birgitt. 2000. Shared Responsibility: Some Aspects of Gender and Authority in Makasar Society. In *Authority and Enterprise among the Peoples of South Sulawesi. Transactions, Traditions, and Texts among the Bugis, Makasarese, and Selayarese,* Eds. R. Tol, K. van Dijk, G. Acciaioli, 143–160. Leiden: KITLV Press.

Rubatscher, Maria Veronika. 1928. Wie aus dem Zwerger Hansele ein Riese wurde. In *Sonderdruck aus den Veröffentlichungen des Museums Ferdinandeum,* 37–44. Innsbruck: Ferdinandeum.

Sambell, K., and L. McDowell. 1998. Assessment & Evaluation. In *Higher Education* 23/4, 391–402.

Schiedeck, Jürgen. 2000. Sprechen. Neues vom pädagogischen Sprachmarkt", In *Erziehung als offene Geschichte. Vom Wissen, Sprechen, Handeln und Hoffen in der Erziehung,* Eds. H. Kupffer, J. Schiedeck, D. Sinhart-Pallin, and M. Stahlmann, 52–76. Weinheim and Basel: Beltz und Deutscher Studienverlag.

Schivelbusch, Wolfang. 2003. *The Culture of Defeat: On National Trauma, Mourning, and Recovery.* Translated from German by Jefferson Chase. New York: Henry Holt and Co.

Schratz, Michael. 2009. 'Lernseits' von Unterricht. Alte Muster, neue Lebenswelten – was für Schulen. *Lernende Schule,* 12 (46–47), 16–21.

Schratz, Michael. 2012. Alle reden von Kompetenz, aber wie!? Sehnsucht nach raschen Lösungen. *Lernende Schule* 15 (58), 17–20.

Schratz, M., J. F. Schwarz, and T. Westfall-Greiter. 2012. *Lernen als bildende Erfahrung. Vignetten in der Praxisforschung.* Innsbruck: Studienverlag.

Schütz, Alfred. 1967. *The Phenomenology of the Social World.* Evanston, IL: Northwestern University Press.

Schützenbund. 2019. *Marketenderinnen. Frauen im Südtiroler Schützenbund.* https://schuet zen.com/organisation/arbeitsgruppen/marketenderinnen/. Accessed 10 November 2019.

Schützenbund. 2020. *Statuten des Südtiroler Schützenbundes.* https://schuetzen.com/organi sation/ssb/statuten/. Accessed 20 January 2020.

Setsche, Michael. 2016. *Wider die Rede vom Postfaktischen. Soziologischer Zwischenruf zur medialen Konstruktion von Wirklichkeit.* https://www.heise.de/tp/features/Wider-die-Rede-vom-Postfaktischen-3562756.html. Accessed 28 December 2016.

Sinhart-Pallin, D., and Stahlmann, M. 2000. Illusionen der Pädagogik. In *Erziehung als offene Geschichte. Vom Wissen, Sprechen, Handeln und Hoffen in der Erziehung,* Eds. H. Kupffer, J. Schiedeck, D. Sinhart-Pallin, and M. Stahlmann, 7–13. Weinheim and Basel: Beltz und Deutscher Studienverlag.

Sohn, W., and Mehrtens, H., Eds. 1999. *Normalität und Abweichung. Studien zur Theorie und Geschichte der Normalisierungsgesellschaft.* Wiesbaden: Westdeutscher Verlag.

SQA. 2017. Pädagogische Diagnostik. *Schulqualität Allgemeinbildung,* 10 October 2017. http://www.sqa.at/pluginfile.php/777/course/section/329/reader_paedagogische_diagnos tik_171010.pdf. Accessed 10 January 2022; the website has been transferred in 2021 to https://www.qms.at/.

Standard. 2014. „Europaweit kaum separate Sprachförderklassen". *Der Standard,* 28 June 2014. https://derstandard.at/2000003615661/Europaweit-kaum-separate-Klassen-fuer-Sprachfoerderung. Accessed 20 April 2017.

Standard. 2016. Plakatkampagne: Van der Bellen setzt auf *Heimat*liebe und Zusammenhalt. *Der Standard,* 21 March 2016. https://derstandard.at/2000033311073/Plakatkampagne-Van-der-Bellen-setzt-auf-Heimatliebe-und-Zusammenhalt. Accessed 8 June 2016.

Stern, Daniel N. 1977. *The first relationship: Infant and Mother.* London: Open Books.

Südtirol News. 2013. Hautfarbe bei den Schützen wird bunter. *Südtirol News,* 25 November 2011. www.suedtirolnews.it/d/artikel/2013/11/25/hautfarbe-bei-den-schuetzen-wird-bunter.html. Not any more accessible, cf. Ein Tirol 2013.

Surowiecki, James. 2004. *The wisdom of crowds.* New York: Doubleday.

Tol, R., K. Van Dijk, and G. Acciaioli, Eds. 2000. *Authority and Enterprise among the Peoples of South Sulawesi. Transactions, Traditions, and Texts among the Bugis, Makasarese, and Selayarese.* Leiden: KITLV Press.

Transit Migration. 2007. *Turbulente Ränder. Neue Perspektiven auf Migration an den Grenzen Europas.* Bielefeld: Transcript Verlag.

Trenker, Luis. 1979. *Alles gut gegangen.* Munich: Bertelsmann.

Tschenett, Stefan. 2004. *Feuernachtsmord. Geschichte eines Todes.* Bozen-Bolzano: stv Verlag.

UNESCO. 2009. *Policy Guidelines on Inclusion in Education.* Paris: United Nations Educational, Scientific and Cultural Organization. https://unesdoc.unesco.org. Accessed 20 March 2020.

UNESCO. 2021. *Reimagining Our Futures Together. A new social contract for education. Report from The International Commission on the Futures of Education.* https://relief web.int/report/world/reimagining-our-futures-together-new-social-contract-education. Accessed 17 december 2021.

United Nations. 2015. *Transforming our World: The 2030 Agenda for Sustainable Development.* https://sustainabledevelopment.un.org/post2015/transformingourworld/public ation. Accessed 20 October 2021.

United Nations. 2016. *Convention on the Rights of Persons with Disabilities* (CRPD). https:// www.un.org/development/desa/disabilities/convention-on-the-rights-of-persons-with-disabilities/convention-on-the-rights-of-persons-with-disabilities-2.html. Accessed 14 February 2019.

Unterrichter, Katharina von. 2007. *Heimat, Migration und Identität. Die Bedeutung von „Heimat" für Menschen mit Migrationshintergrund am Beispiel der Südtiroler Heimatfernen in Niedersachsen.* Vienna: Institut für Kultur- und Sozialanthropologie.

Vattimo, Gianni. 1987. "Verwindung": Nihilism and the Postmodern. *Philosophy. Contempory Italian Thougt* Vol. 16/2 (53), pp. 7–17. https://doi.org/10.2307/3685157.

Vattimo, Gianni. 1997. *Beyond Interpretation: The Meaning of Hermeneutics for Philosophy.* Palo Alto: Stanford University Press.

Vattimo, Gianni. 2013. *Weak Thought,* Ed. Pier Aldo Rovatti. Translated by Peter Carravetta. New York: SUNY Press.

Vattimo, Gianni. 2019. *Beyond the Subject: Nietzsche, Heidegger, and Hermeneutics.* Translated by Peter Carravetta. New York: SUNY Press.

Vetter, A., and B. Best. 2015. Konvivialität und Degrowth. Zur Rolle von Technologie in der Gesellschaft. In *Konvivialismus. Eine Debatte,* Eds. F. Adloff, and V. Heins, 101–112. Bielefeld: Transcript.

Vogel, Dagmar. 2019. *Habitusreflexive Beratung im Kontext von Schule. Ein Weg zu mehr Bildungsgerechtigkeit.* Wiesbaden: Springer VS.

Volkan, V.D., and G. Ast. 1994. *Spektrum des Narzissmus. Eine klinische Studie des gesunden Narzissmus, des narzisstischen-masochistischen Charakters, der narzisstischen Persönlichkeitsorganisation, des malignen Narzissmus und des erfolgreichen Narzissmus.* Göttingen-Zurich: Vandenhoeck & Ruprecht.

Volkan, Vamik D. 2003. Large-group identity: Border psychology and related societal processes. In *Mind and Human Interaction* 13, pp. 49–76.

Voß, Heinz-Jürgen. 2010. *Making Sex Revisited: Dekonstruktion des Geschlechts aus biologisch-medizinischer Perspektive.* Bielefeld: Transcript.

Waldenfels, Bernhard. 2000. *Das leibliche Selbst.* Frankfurt Main: Suhrkamp.

Waldenfels, Bernhard. 2004a. *Phänomenologie der Aufmerksamkeit.* Frankfurt Main: Suhrkamp.

Waldenfels, Bernhard. 2004b. Bodily experience between selfhood and otherness. *Phenomenology and the Cognitive Sciences* 3, 235–248 https://doi.org/10.1023/B:PHEN.000 0049305.92374.0b. Accessed 10 march 2019.

Waldenfels, Bernhard. 2009. Lehren und Lernen im Wirkungsfeld der Aufmerksamkeit. In *Umlernen. Festschrift für Käte Meyer-Drawe,* Eds. N. Ricken, H. Röhr, J. Ruhloff, and K. Schaller, 23–33. Munich: Wilhelm Fink.

Weizsäcker, E. U. von, and A. Wijkman, Eds. 2017. *Come On! Capitalism, Short-termism, Population and the Destruction of the Planet.* Heidelberg: Springer.

Welsch, Wolfgang. 1999. Transculturality – the Puzzling Form of Cultures Today. In *Spaces of Culture: City, Nation, World*, Eds. M. Featherstone and S. Lash, 194–213. London: Sage

Werning, Rolf. 2016. Lernen. In *Handbuch Inklusion und Sonderpädagogik*, Eds. I. Hedderich, G. Biewer, J. Hollenweger, and R. Markowetz, 229–233. Bad Heilbrunn: Klinkhardt.

Winnicott, Donald W. 1965. *Maturational Processes and the Facilitating Environment: Studies in the Theory of Emotional Development*. London: Hogarth Press.

Wintersteiner, Werner. 2021. *Die Welt neu denken lernen. Plädoyer für eine planetare Politik. Lehren aus Corona und anderen existentiellen Krisen*, Ed. H.K. Peterlini. Bielefeld: transcript.

Wuttig, Bettina. 2016. *Das traumatisierte Subjekt. Geschlecht – Körper – Soziale Praxis. Eine gendertheoretische Begründung der Soma Studies*. Bielefeld: Transcript.

WWF. 2016. *Living Planet Report 2016. Risk and resilience in a new era*. Gland: WWF International. http://awsassets.panda.org/downloads/lpr_2016_full_report_low_res.pdf. Accessed 17 December 2019.

Xypolia, Ilia. 2016. Divide et Impera: Vertical and Horizontal Dimensions of British Imperialism. *Critique: Journal of Socialist Theory* 44 (3), 221–231. https://doi.org/10.1080/030 17605.2016.1199629. Accessed 20 May 2020.

Yildiz, E., M. Hill, Eds. 2014. *Nach der Migration: Postmigrantische Perspektiven jenseits der Parallelgesellschaft*. Bielefeld: Transcript.

Zirfas, Jörg. 2010. Identität in der Moderne. Eine Einleitung. In *Schlüsselwerke der Identitätsforschung*, Eds. Jörissen, B., and J. Zirfas, 9–18. Wiesbaden: VS.

Printed in the United States
by Baker & Taylor Publisher Services